A CULINARY JOURNEY OF GRATITUDE COOKBOOK

PHOTOGRAPHY BY
DAVE PERRY & BIA RODRIGUES PERRY

thebookchief.com

PUBLISHED BY THE BOOK CHIEF PUBLISHING HOUSE 2024
(A TRADEMARK UNDER LYDIAN GROUP LTD)
Suite 2a, Blackthorn House, St Paul's Square,
Birmingham, B3 1Rl

WWW.THEBOOKCHIEF.COM

COPYRIGHT © 2024 BY BIA RODRIGUES PERRY

All rights reserved. No part of this book may be reproduced, stored in a retrieval system, or transmitted in any form or by any means, electronic, mechanical, photocopying, recording, public performances or otherwise, without written permission of Bia Rodrigues Perry, except for brief quotations embodied in critical articles or reviews. The book is for personal and commercial use by the author Bia Rodrigues Perry.

The right of Bia Rodrigues Perry to be identified as the author of this work has been asserted in accordance with sections 77 and 78 of the copyright designs and patents act 1988.

ISBN: 978-1-06869810-1

BOOK COVER DESIGN: Bia Rodrigues Perry / Abu Bakar Javed
EDITING: Sharon Brown
PUBLISHING: Sharon Brown

PUBLISHED BY THE BOOK CHIEF PUBLISHING HOUSE

OLIVER PERRY
14 AUGUST 1974 - 15 APRIL 2021

I dedicate this book in loving memory of my dear brother-in-law, Oli Perry.

In 2021, we tragically lost Oli to his battles with mental health. Oli was not only a dedicated Personal Trainer and Online Coach but also a two-time WNBF World Champion and two-time British Champion. His unwavering passion for his craft and infectious sense of humour had a lasting impact on everyone who knew him.

The children still treasure his collection of joke books, and we all admired his commitment to his profession. I can't help but smile as I remember him constantly asking for recipes—no matter how many times I shared them. Eventually, I'd laugh and say, "Shall I just make it for you?"

To which he'd always reply, "Yeah, that might be best—I wouldn't want to get it wrong and disappoint you! Ha ha ha."

Oli, your warmth, humour, and spirit will forever be loved and remembered.

With heartfelt love,
Bia

TABLE OF CONTENTS

Introduction 6
Bia's Kitchen Show Guests' Recipes 7
Pastel De Carne By Tropical Brazilian Foods 8
3 For 1 - Quinoa And Grilled Chicken Salad By Nikki Ryan 11
Parma Ham Carpaccio And Bia's Salad Dressing By Bia Rodrigues................13
Paprika Club 15
Keema Curry By Paprika Club 16
Kurzi Lamb By Paprika Club 17
Chocolate Chip Cookies By Emi's Little Bakery...19
Tamarind & Caramel Fried Chicken By chef Ash Faria - Magic Wingdom......22
Tamarind And Caramel Sauce By chef Ash Faria - Magic Wingdom...............24
Smacked Cucumber Salad By chef Ash Faria - Magic Wingdom....................25
Homewrecker Chicken Tetrazzini 28
Salmon Poke Bowl By A Good Catch 31
Confit Duck Rissole With Crispy Duck Skin, Horseradish Gel & Pickled Courgettes,
By Sophie Hyam 34
Supreme Of Masala Chicken, Crushed Sweetcorn & Pickled Onions
By Scratch Cook Social 38
Thai Salmon Red Curry By Mari-Carmen 44
Plain Scones By Chef Paul From Whittle's Restaurant 49
Wild Garlic Salsa Verde And Truffled Parmesan Custard
By Chef Greg Newman - The Woodsman 53
Focaccia By Chef Greg Newman - The Woodsman..54
Asparagus By chef Greg Newman - The Woodsman 56
Honey, Orange And Pistachio Cake By Sophie Page 57
Hazelnut Praline Bon Bons By Chef Michael Scott 61
Cocktails 65
Elderflower Fizz. By David From The Warwickshire Gin Co. 66
Passion Fruit Martini/Pornstar By David From The Warwickshire Gin Co. 67
Espresso Martini By Steve From That Gin Company 69
Passion Fruit Martini By That Gin Company 71
Neopolitan Pizza By Chef Marius From The Leopard Spots 73
Duck Hagu With Fettuccine By Chef Martin From Tavóla 78

4

Octopus Lasagne By Chef Angelo - Rodizio Rico	83
Pancetta And Mushroom Risotto By Chef Angelo - Rodizio Rico	88
Bia's Twist On Her Favourite Recipes	92
Bia's Feijoada Black Bean Stew	93
Bia's Brazilian Seafood Curry	96
Chilli Con Carne	98
Beef Stroganoff	100
Bia's Fusion Chicken Curry	103
Green Pea Soup With Crispy Bacon	106
Brigadeiro - Brazilian Chocolate Truffles	108
Passion Fruit Mousse And Dark Chocolate Cheesecake	111
Bia's Chocolate & Pistachio Tower Cake	114
Kibe	118
Pão De Queijo - Cheese Bread	121
Steak In The Pan	124
Brazilian Bbq	126
Meats	128
Molho A Campanha - Vinaigrette Salsa	129
Farofa - Yucca Flour	130
Garlic Rice	131
Beef Carpaccio	132
Detox Oven Omelette	135
Brazilian Salt Cod Pie - Bacalhau Espiritual	139
Creamy Risotto With Roast Honey Beetroot	142
Beetroot With Honey And Caramelised Nuts	143
Risotto	144
Beetroot And Caramelised Nuts	145
Bia's Christmas Tree	146
About Our Guests	147
The Team That Makes It Happen	170
Why Choose Mind As Our Charity?	171
Sponsors	172

INTRODUCTION

"A Culinary Journey of Gratitude"
A Tribute to Passion for Food and People

This book pays tribute to the love of independent food and the people behind it. The author shares her challenging experience of moving to England, leaving behind her career, family, and friends. She faced pregnancy-related issues and struggled to find her identity, but her passion for food and faith helped her overcome this difficult period.

In 2023, she launched the Bia's Kitchen Show, an online TV series that supports the independent food industry in her Warwickshire community by sharing real people's stories and featuring their signature dishes.

The book showcases remarkable independent chefs, cooks, restaurants, nutritionists, and more, highlighting their unique dishes that can be easily recreated at home. It also includes some of the author's favourite recipes.

Furthermore, all profits from this book will be donated to a charity that help women with postnatal depression and men with mental health issues. This is a way to share love and support for one another."

BIA'S KITCHEN SHOW GUESTS' RECIPES

TROPICAL BRAZILIAN FOODS
PASTEL DE CARNE

Exploring the Flavours of Tropical Brazilian Foods with Chef Tatiana

Tropical Brazilian Foods is run by Tatiana, a self-taught Brazilian chef who serves a diverse range of delectable dishes at her food truck in Coventry.

Tatiana was the inaugural guest on our show, and we had the pleasure of preparing one of my favourite Brazilian street foods, pastel de carne, together. It was a delightful experience that conjured fond memories of the lively streets of Rio de Janeiro.

INGREDIENTS

For homemade dough (or you can buy ready-made dough from Jaboque Foods UK)

Dough (basic recipe):
- 500g of plain flour
- 75ml of oil
- 250ml of water
- 1/2 dessert spoonful of salt

Dough Preparation:
Mix all ingredients and work the dough until combined. Then, stretch the dough using a pasta cylinder. This can also be done by hand, but getting a fine dough at the end will be much more difficult.

Mince Beef (filling):
- 500g of minced beef
- 3 cloves of garlic (finely chopped)
- 1 medium onion (finely chopped)
- 1 tsp of salt
- 1 tsp of ground black pepper
- 2 tsp of sweet paprika
- 1 tsp of ground cumin
- 1 handful of fresh parsley (finely chopped)

PREPARATION:

1. Preheat the pan and add the minced beef; keep stirring until the meat changes colour.
2. Add the chopped ingredients (onion and garlic), and continually stir so the meat doesn't burn and stick to the pan.
3. Add all the spices, let it cook for about 5 minutes, and add the fresh parsley.
4. Fill the pastry in whatever shape you prefer as long as it is sealed, and then deep fry.
5. Enjoy for dinner or as a snack for children.

TIPS:

If you like a sweet version, fill it with banana and cinnamon. Or, if you prefer, choose your favourite veggies and mix them with cheese, making another delicious filling.

3 FOR 1 - QUINOA AND GRILLED CHICKEN SALAD

BY NIKKI RYAN / WEIGHT LOSS COACH

LET'S COOK

QUINOA:

- 200g (1 cup) Quinoa
- 1 vegetable stock cube
- 400ml water

PREPARATION:

Put the quinoa into a saucepan. Pour in 400ml of cold water and add the vegetable stock. Bring to a boil over medium heat, then reduce to a simmer over low heat and cook gently for 10-15 minutes until the quinoa is fully cooked and all the water has been absorbed. Remove from the heat, cover with a lid, and let it sit for 5 minutes. Fluff it up with a fork, and it's ready to serve.

CRISPY SALAD:

In a large bowl, combine:

- 1 bag of crispy salad
- 2 grilled chicken breasts (or meat from a whole roasted chicken, sliced)
- 1/4 red onion, peeled and sliced
- 10 cherry tomatoes, halved
- 200g cooked sweetcorn (canned is fine)
- Any of your favourite vegetables (e.g. raw broccoli)
- Optional: feta cheese

Mix the ingredients well and drizzle Bia's Salad Dressing on top of the salad.

PARMA HAM CARPACCIO AND BIA'S SALAD DRESSING

BY NIKKI RYAN / WEIGHT LOSS COACH

LET'S COOK

BIA'S SALAD DRESSING:

In an empty pot with a lid, add 50ml of olive oil, 1 clove of garlic (peeled and minced), 3 tbsp of balsamic vinegar, 30ml of double cream (for a vegan option, use a suitable alternative), 1 tbsp of honey, and a pinch of salt. Put the lid on the pot and shake it well.

PARMA HAM CARPACCIO:

Place 100g of Parma ham on a wooden chopping board or plate. Add a bag of rocket salad, a handful of cherry tomatoes (sliced in half), five thinly sliced strawberries, 2 tbsp of capers, 2 tbsp of balsamic vinegar, 1 tbsp of honey, and three drizzles of olive oil. A pinch of flake salt and black pepper are optional. Ready to serve, as simple as that!

PAPRIKA CLUB

KEEMA CURRY

INGREDIENTS

- 1 large onion chopped
- 2 garlic cloves chopped
- 4cm piece ginger grated
- 2 green chillies
- 3 tbsp oil
- 500g lamb mince
- 2 tbsp garam masala
- 2 medium fresh tomatoes
- 2 tbsp of mixed paprika powder (haldi powder, curry powder, jeera powder)
- 1 small bunch coriander chopped

PREPARATION:

STEP 1
Chop the onion, garlic, ginger, and chillies and blend them in a food processor. Heat the oil in a large frying pan and fry the mixture until it becomes fragrant. Add the mince and fry until it begins to brown, stirring to break up any lumps.

STEP 2
Add the spices and fry for 1 minute. Add the tomatoes and bring to a simmer, cook for 1 minute, add a pinch of salt and a good grind of black pepper, add a splash of water if necessary, and cook the mixture for 30 minutes. Stir in the coriander.

KURZI LAMB

INGREDIENTS

- 1 tsp of ginger paste
- 1 tsp of garlic paste
- 4 tsp of olive oil
- 2 tsp paprika mix spices (haldi powder, curry powder, jeera powder and Danja powder mixed in equal amounts)
- 1 tsp of dry methi (fenugreek)
- 1/2 tsp salt
- 1 tsp of Kashmiri paste
- 1 tsp of tikka paste
- 1 tsp tandoori paste
- 3 tsp of yoghurt
- Juice of 1 lemon
- 10 garlic gloves

PAPRIKA'S TEAM

PREPARATION:

1. Tip the fragrant spice mix into a bowl and stir well to combine.
2. Cut deep slits into the lamb and rub the spice mixture all over the meat.
3. Make sure to push some of the mixture into the slits and insert the garlic cloves into the slits.
4. Place the lamb into a large sealable bag and refrigerate overnight.
5. Preheat the oven to gas mark 4 (180°C or 160°C fan).
6. Place the lamb on a roasting rack in a roasting tin, cover with foil, and bake for 45 minutes to 1 hour.
7. Remove the foil and cook for an additional 30–45 minutes.
8. Cover the lamb from the oven, and let it rest for 30 minutes before serving.

EMI'S LITTLE BAKERY
CHOCOLATE CHIP COOKIES

INGREDIENTS

- 120g Salted Butter
- 40g White Granulated Sugar
- 160g Soft Brown Sugar
- 1 Egg
- 2 tsp Vanilla Extract
- 1 tsp Bicarbonate of Soda
- 1 tsp Baking Powder
- 300g Plain Flour
- 250g Belgian Chocolate Chips

METHOD

1. Combine the butter and sugar, then add the vanilla and egg and whisk to combine.
2. Add flour, Bicarbonate of Soda and Baking Powder and mix until combined.
3. Fold the chocolate chips and roll them into small balls (around 120g).
4. Seal in a bag and chill in the fridge overnight. Heat oven to 160°C fan or 180°C top/bottom heat. Place 4 chilled dough balls on a lined baking tray and bake for 12 minutes.
5. Take out of the oven and leave to cool on the tray before transferring to a rack

AND JUST LIKE THAT THEY ARE READY!

ENJOY IT!

TAMARIND & CARAMEL FRIED CHICKEN

BY CHEF ASHLEY FARIA
THE MAGIC WINGDOM

INGREDIENTS

For the chicken:

- 500g plain flour
- 100g cornflour
- 1 tbsp Oregano
- 1 tbsp garlic powder
- 100g crushed cornflakes
- Salt
- Pepper
- Spices of your choice (chilli, paprika etc)

METHOD

1. Mix all the ingredients,
2. Mix 1/3rd of the mix with 300ml water to create a batter (thick). Add more water if you like a lighter batter to your desired consistency.
3. To prepare the chicken, place the chicken into the batter mix and then into the flour.
4. When all your chicken is coated, put it into a fryer at 175°C for 7-8 minutes until golden and the core temperature has exceeded 75°C

TAMARIND AND CARAMEL SAUCE

INGREDIENTS

Serves 4 (Vegetarian)
For the Caramel:

- 150g Brown sugar
- 75g Water

In a heavy-bottomed pan, combine the brown sugar and water, then bring to a boil. Be careful not to let it boil over. Once the caramel reaches a rolling boil, add the following ingredients, reduce the heat, and stir until the sauce coats the back of a spoon.

To Finish the Sauce:

- 30g Water
- 3 tbsp Tamarind pulp
- 1½ tbsp Fish sauce
- Juice of a quarter lime
- 5g Cornflour

Combine the ingredients and whisk until smooth. After mixing into the caramel, add a whole red chilli.

To Garnish the Sauce:

- 8 Mint leaves
- 8 Basil leaves
- 30g Crispy onions

Finely slice the mint and basil leaves, sprinkle them over the sauce, and top with crispy onions

This recipe can easily be made vegan by substituting the fish sauce with a tablespoon of red miso paste.

Recipe Credit:
Frankie Griffiths
The Magic Wingdom

SMACKED CUCUMBER SALAD

SMACKED CUCUMBER SALAD

6 portions (Vegan)

A lightly spiced, refreshing cucumber salad, perfect as a snack or palate cleanser.

Prep time: 20 minutes

You will need:

- A blender
- Knife
- Mixing bowl

INGREDIENTS

- Cucumber (roughly 2 cucumbers 600g)
- Sea Salt 15g (a good pinch)

PREPARATION:

Start preparing the cucumbers by trimming the ends. Place each cucumber on a wooden chopping board. Smash it gently along the length using a chef's knife and the palm of your hand or rolling pin until it splits. Slice the cucumbers in 1-2 inch diagonal slices. Using a spoon, scoop the cucumber pieces into a medium bowl. Sprinkle with a pinch of sea salt flakes. Set aside for 15 minutes. Then drain off the excess liquid.

For the Dressing;
Place the following ingredients into a small blender and blend until smooth.

- 18g Maggie
- 4g Toasted Sesame oil
- 30g Soy Sauce
- 60g Rice wine vinegar
- 20g Ginger (no need to peel, chop it up!)
- 20g Garlic
- 10g Brown sugar
- 3g Gochugaru

Allow the sauce to sit for 10 minutes to mellow the garlic before use. This dressing will keep in the fridge for three days.

Combine the dressing and drained cucumber in a mixing bowl.

GARNISH

- Toasted Sesame seeds -5g

Dig in!

Recipe credit to
Ashley Faria
The Magic Wingdom

HOMEWRECKER CHICKEN TETRAZZINI

INGREDIENTS

Serves: 4-6

- 2 chicken breasts
- 200-250g fresh tagliatelle
- 50g butter
- 30g flour
- 400g single cream
- 50g sherry
- 1 tsp garlic granules
- 5 good-sized mushrooms
- 4 spring onions
- 50g piquante peppers
- 2 tsp Homewrecker cheese

Optional:
Grated Grana Padano or Parmesan for topping

METHOD

1. Preheat the oven to 180°C (350°F). Marinate the chicken breasts with truffle oil (or your preferred oil), sea salt, and cracked black pepper.
2. Bake the chicken for 30 minutes.
3. Once cooked, set aside to cool, then dice the chicken. Note: Leftover roast chicken can also be used.
4. Grease a 9 x 13-inch casserole dish with a bit of melted butter and set aside.
5. Following the time indicated on the packaging. Boil the tagliatelle (or spaghetti) in salted boiling water with oil.
6. Once cooked, drain the pasta and place it in the buttered casserole dish. Set aside.
7. Chop the mushrooms and spring onions.
8. Sauté them in a pan with butter until soft.
9. Remove from heat and mix in the piquante peppers and Homewrecker cheese.
10. Add the diced chicken to the vegetable mixture and mix well.
11. Melt the butter in a saucepan.
12. Once melted, whisk in the flour until well mixed but not brown.
13. Whisk in the sherry, then slowly add the cream while whisking continuously.
14. Once the sauce thickens, mix in the garlic granules.
15. Add the chicken and vegetable mixture to the sauce and mix well.
16. Pour the sauce mixture over the pasta in the casserole dish.
17. Top with grated Grana Padano or Parmesan if desired.
18. Bake for 20-25 minutes or until the top is golden brown.

SALMON POKE BOWL WITH SUSHI RICE AND FRESH TOPPINGS

BY A GOOD CATCH

INGREDIENTS

Servings: 2

For the sushi rice:
- 1 cup sushi rice
- One ¼ cup of water
- 2 tbsp rice vinegar
- 1 tbsp sugar
- 1/2 tsp salt

For the salmon:
- 250g fresh salmon, diced
- 1 tbsp soy sauce
- 1 tsp sesame oil (optional)
- 1 tsp rice vinegar
- 1 tsp honey or sugar (optional)

Toppings:
- 4-5 gherkins, sliced
- 2 tbsp sushi ginger
- 1/4 cup shredded red cabbage
- 1/2 mango, diced
- 1/2 cup edamame (shelled)
- 1/2 avocado, sliced or diced
- Two spring onions, sliced
- 4-5 radishes, thinly sliced

For the dressing:
- 2 tbsp soy sauce
- 1 tbsp mayonnaise (preferably Japanese Kewpie mayo)
- 1 tsp sriracha (optional for a spicy kick)
- Sesame seeds (for garnish)

METHOD

Prepare the sushi rice:
1. Rinse the sushi rice under cold water until the water runs clear.
2. Combine the rinsed rice and water in a pot. Bring to a boil, then reduce to a simmer and cover with a lid. Cook for about 18-20 minutes, or until the water is absorbed and the rice is tender.
3. Once cooked, remove from heat and let it rest for 10 minutes.
4. Mix the rice vinegar, sugar, and salt in a small bowl. Stir until dissolved, then gently fold it into the rice. Let the rice cool to room temperature.

Prepare the salmon:
1. Combine the soy sauce, sesame oil, rice vinegar, and honey (if using) in a small bowl.
2. Add the diced salmon to the marinade and let it sit in the fridge for at least 10 minutes while you prepare the other ingredients.

Prepare the toppings:
1. Slice the gherkins, sushi ginger, radish, and spring onion.
2. Dice the mango and avocado, and shred the red cabbage.
3. If needed, cook the edamame in boiling water for 2-3 minutes, then drain and set aside.

Assemble the poke bowl:
1. You can start by placing a generous portion of sushi rice at the bottom of each bowl.
2. Arrange the marinated salmon and toppings (gherkin, sushi ginger, red cabbage, mango, edamame, avocado, spring onion, and radish) in sections on top of the rice.

Add the dressing:
Drizzle soy sauce and mayonnaise over the bowl. If you prefer a spicy touch, mix the mayo with sriracha before adding it.
Sprinkle with sesame seeds for garnish.

Serve and enjoy!
You can also serve it with extra soy sauce or wasabi on the side if desired.
This poke bowl is fresh, colourful, and packed with flavour.

CONFIT DUCK RISSOLE WITH CRISPY DUCK SKIN, HORSERADISH GEL & PICKLED COURGETTES.

BY THE AWARD-WINNING CHEF SOPHIE HYAM

INGREDIENTS

Serves 4 as a starter

Serves 2 as a main

- Confit Duck Legs
- 3 juniper berries
- 50g flaky sea salt
- 4 duck and thigh joints
- 1 small bunch thyme
- 1 twig of rosemary
- 4 garlic cloves
- 500g goose or duck fat, or enough to submerge the duck legs
- 2 bay leaves
- 1 tsp black peppercorns
- 4 beaten eggs (1 for binding & 3 for panning the rissoles) Handful of chopped fresh parsley
- 1 tsp cumin powder
- 150g plain flour (40g for duck mix and 110g for pane)
- 250g panko breadcrumbs
- 500ml vegetable oil Handful of fresh pea shoots to garnish

METHOD

STEP 1
Crush the juniper berries and mix with the salt the day before cooking. Rub the mixture over the duck, scatter with thyme, rosemary and sliced garlic and chill for 24 hrs, turning two or three times as they marinate.

STEP 2
The next day, heat the oven to 150°C/130°C fan/ gas 2. Wipe the duck with kitchen paper and pat dry, but don't wash off the marinade.

STEP 3
Put the duck in a cast-iron casserole dish and cover with the goose fat or duck fat. Add the bay leaves and peppercorns and cook for about 2½ hrs, or until the meat almost falls away from the bone. Allow to cool. Remove the duck skin carefully from the leg and put it to one side. Pull all the duck meat off the legs, shred it, and put it in a mixing bowl. Add one beaten egg and 40 grams of plain flour. Season with a good pinch of sea salt, one teaspoon of cumin powder and a good grind of black pepper. Add the chopped parsley and taste to check the seasoning. Roll into rissoles and pop in the fridge for 30 minutes.

STEP 4
Place the duck skin in a preheated oven at 180°C for 35 minutes until golden and crispy. Set to one side.

STEP 5
Assemble the duck rissoles. Place three bowls in a row: Flour, beaten eggs, and panko breadcrumbs. Place each rissole in the flour, egg, and breadcrumbs to form an outer coating. When all coated, pop back in the fridge until ready to fry.

METHOD CONT'D

STEP 6
Bring the vegetable oil to medium heat and put the rissoles in the pan. Don't overcrowd the pan, as this will lower the temperature of the oil, and it won't crisp up. Alternatively, they can be cooked in an air fryer - 180°C for 18 mins. Once golden brown, pop onto a kitchen towel to drain the oil and prepare to plate up.

- Add 400g of Duck Gravy
- 1lb of chicken wings
- Vegetable oil
- 4 carrots, peeled and chopped into 3cm pieces
- 4 celery stalks, chopped into 3cm pieces
- 1 onion, peeled and chopped into 3cm pieces
- 1 garlic bulb, cut in half
- 150g clear honey
- 4 cloves
- 2 litres of brown chicken stock
- 500g unsalted butter
- 1 lemon, juice only

(Method for the gravy) Preheat the oven to 180°C.

1. Chop the chicken bones, place them into a flameproof roasting tin, and roast in the oven for 20-30 minutes or until golden brown.
2. Heat the oil in a large saucepan and fry the carrots for 5-10 minutes or until almost black. Add the celery, onion and garlic and fry for 4-5 minutes.
3. Once again, allow for blackening/char.
4. Remove the bones from the tray and add to the saucepan.
5. Drain off any excess fat from the pan, add the honey and cloves and cook until the honey caramelises.
6. Deglaze the pan with a little of the chicken stock.
7. Add the remaining chicken stock to the pan and cook until the volume of the liquid has reduced by half.
8. Strain the mixture through a muslin cloth and skim off any excess fat.
9. Measure one litre of the gravy into a saucepan, add the butter and return the pan to the heat.
10. Cook the sauce until emulsified and reduced slightly.
11. Season, to taste, and add lemon juice.
12. Place on both of your plates.
13. Top with the rissoles.
14. Garnish with the crispy duck skin and pea shoots.

SUPREME OF MASALA CHICKEN, CRUSHED SWEETCORN & PICKLED ONIONS

BY SCRATCH COOK SOCIAL

INGREDIENTS

For the Crushed Sweetcorn

- 350g of sweetcorn, lightly blitzed in a food processor, maintaining a rough texture 70g of banana shallots, finely chopped (approx. two medium-sized shallots)
- 7g crushed garlic clove
- 7g finely chopped ginger + 5g for finishing the dish
- 2 tbsp Coconut oil
- 10 curry leaves
- 1 tsp mustard seeds
- ¼ tbsp turmeric powder
- 2 tsp coriander-cumin powder
- 1 tsp Maldon salt
- 400ml of coconut milk (1 tin)
- Squeeze of lemon
- Fresh Coriander, chopped

For the Chicken Marinade

- 4 supreme of chicken, skin off
- 8 cloves of garlic (approximately 30g)
- 4 fresh green chillis
- 30g fresh ginger
- 1 tbsp of Scratch Cook Social's Handcrafted Tandoori Masala
- 2 tbsp Malden Salt
- Juice of half a lemon
- 2 tbsp of light olive oil

FOR THE CRUSHED SWEETCORN

METHOD

1. Heat the oil in a pan; add mustard seeds and curry leaves.
2. Once seeds begin to pop, add the shallots and turn down to a low-medium heat to avoid burning.
3. Slowly cook the shallots until softened, then add the garlic and the 7 grams of ginger and cook for 2 minutes.
4. Add the sweetcorn, turmeric, coriander, cumin and salt.
5. Stir and cook for 5 minutes.
6. Add the coconut milk and leave to simmer for 15 minutes with the lid on at low-medium heat.
7. Check the salt and add a touch more if needed.
8. Before serving, add the remaining 5 grams of ginger, a squeeze of lemon and chopped coriander.

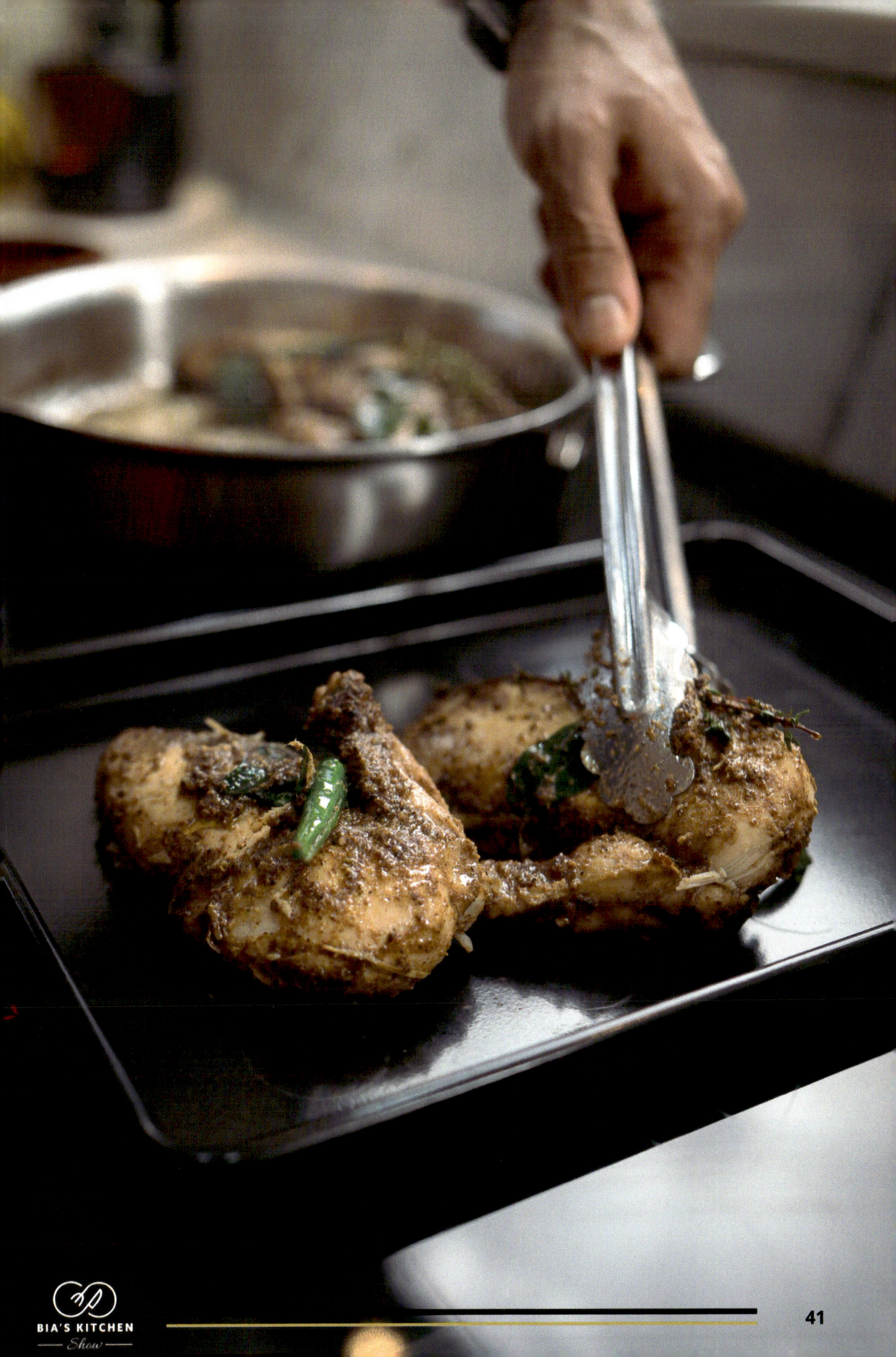

PICKLED ONIONS

METHOD

Mince the garlic, ginger and chilli and mix the other ingredients to create a paste.

Coat the chicken pieces fully and let them rest in the fridge overnight if possible, or for at least 3 hours if you're short on time.

In the meantime, cook your chicken, you'll need the following ingredients:

- 2-3 tbsp light olive oil
- 10 curry leaves
- 3 sprigs of thyme
- 50g unsalted butter
- 2 fresh green chillies, sliced lengthways
- 15g ginger, cut into matchstick-size pieces

Pre-heat a fan oven to 180°C.

Heat the oil in a large frying pan.

Add the chicken and sear for 2-3 minutes on each side.

Add the butter, and when melted, add the curry leaves, ginger, thyme and chillis.

Baste the chicken until a light brown colour is reached.

Transfer the chicken onto a baking tray and pour over the cooking juices, placing the ginger, thyme and chillis on top of each chicken breast.

Cook in the oven for 10 mins.

FOR THE CHICKEN

METHOD

- While the chicken is in the oven, prepare your pickled onions;
- 1 red onion, cut in half from top to bottom and finely sliced.
- Bunch of fresh coriander roughly chopped, including the stalks.
- 1 tsp Maldon salt
- 1 tbsp red wine vinegar
- Squeeze of lemon juice
- Rub the salt, vinegar, and lemon juice into the sliced onions.
- Add the chopped coriander.
- Rest for 10 minutes.
- Carve your chicken, place it on a bed of sweetcorn, and finish with the pickled onions.
- Enjoy it!

THAI SALMON RED CURRY

BY MARI-CARMEN

INGREDIENTS

- 200g salmon fillet
- 1 chopped red onion
- 400g coconut milk or cream
- 200g sugar snaps
- 3 cloves of crushed garlic
- 1 tbsp fish sauce
- 5 chopped lime leaves
- 1 tsp coconut oil
- 30g Thai red curry paste
- 1 handful of fresh chopped coriander

SIDE
BROWN RICE

- 150g brown rice
- 300ml of water
- Boil for approximately 25 minutes or until it is cooked through.

BIA'S TIP: After the rice is cooked, add two tablespoons of olive oil and one clove of crushed garlic to add more flavour.

METHOD

STEP 1

Heat the oil in a large pan. Add the onion and garlic and cook gently for about 5 minutes until softened. Pour in the coconut milk and bring to a boil, along with the curry paste and fish sauce. Reduce to a simmer and add lime leaves. Add the salmon chunks for 3 minutes, then add the sugar snaps.

STEP 2

Leave to simmer gently for 3 minutes until the fish flakes easily and the beans are tender. Scatter with the coriander and serve with plain brown rice.

HOMEMADE THAI RED CURRY PASTE (OPTIONAL)

INGREDIENTS

- 1 shallot, chopped, or 1/4 cup purple onion
- 1 stalk fresh lemongrass, minced, or 3 tbsp prepared lemongrass paste
- 1 to 2 red chillies, or 1/2 to 1 tsp cayenne pepper, or 2 to 3 tsp Thai chilli sauce
- 4 cloves garlic
- 1 thumb-size piece of galangal, or ginger, peeled and sliced
- 2 tbsp tomato ketchup
- 2 tbsp fish sauce or 2 tbsp soy sauce plus salt to taste
- 2 tbsp freshly squeezed lime juice
- 1 1/2 to 2 tbsp Latin chili powder, or 1 to 2 tsp Asian chili powder
- 1 to 3 tbsp coconut milk for desired consistency
- 1 tsp shrimp paste, or 1 tsp Thai golden mountain sauce
- 1 tsp sugar
- 1 tsp ground cumin
- 3/4 tsp ground coriander
- 1/4 tsp ground white pepper
- 1/4 tsp ground cinnamon or 1 cinnamon stick, optional

Place all ingredients in a food processor or blender and process well to create a fragrant Thai red curry paste. If it is too thick, add a bit of coconut milk to help blend the ingredients.
It will taste very strong but mellow when you add your curry ingredients and coconut milk.

Once cooked, it will also turn a stronger red colour. You can use it immediately or store it in an airtight container in the refrigerator for up to one week.

The paste can also be frozen for up to six months.

48

PLAIN SCONES

BY CHEF PAUL FROM WHITTLE'S RESTAURANT

Scones are a star of every English afternoon tea; this one is simple and delicious.

And this amazing recipe is made by Chef Paul Worthy.

Today, we're baking up a batch of delicious homemade scones that are perfect for breakfast, afternoon tea, or a cosy snack. These scones are light, fluffy, and super easy to make, even if you're a beginner in the kitchen!

INGREDIENTS

Plain Scones

- 450g plain flour plus extra for dusting the table.
- 75g salted butter plus extra for glazing tops.
- 28g baking powder.
- 50g caster sugar plus extra for glazing tops.
- 250ml milk.
- This should yield around 15 scones.

BIA'S TIP: In my recipe, I include 1 tablespoon of vanilla paste and the zest of 1 lemon for extra flavour.

METHOD

STEP 1

Heat the oven to 220°C/200°C fan/gas 7. Add the self-raising flour, salt, and baking powder to a large bowl and mix well.

STEP 2

Add the butter, then rub in with your fingers until the mixture resembles fine crumbs. Stir in the caster sugar.

STEP 3

Put the milk into a jug and heat in the microwave for about 30 seconds until warm but not hot.

OPTIONAL: Add the vanilla extract and a squeeze of lemon juice, then set aside for a moment.

STEP 4

Make a well in the dry mix, then add the liquid and combine it quickly with a cutlery knife – it will seem pretty wet at first.

STEP 5

Scatter some flour onto the work surface and tip the dough out. Dredge the dough and your hands with a little more flour, then fold the dough over 2-3 times until it's a little smoother. Pat into a round about 4cm deep. Take a 5cm cutter (smooth-edged cutters tend to cut more cleanly, giving a better rise) and dip it into some flour. Plunge into the dough, then repeat until you have your scones. You may need to press what's left of the dough back into a round to cut out some more.

STEP 6

Brush the tops with a beaten egg, or use butter if you are allergic to eggs, then carefully arrange on the hot baking tray. Bake for 10 mins until risen and golden on the top. Eat just warm or cold on the day of baking, generously topped with jam and clotted cream. If freezing, freeze once cool. Defrost, then put in a low oven (about 160°C/140°C fan/gas 3) for a few minutes to refresh. Enjoy it with a cup of tea!

THE WOODSMAN

BY CHEF GREG NEWMAN

WILD GARLIC SALSA VERDE AND TRUFFLED PARMESAN CUSTARD

WILD GARLIC SALSA VERDE

INGREDIENTS

- 4 eggs – boiled for 9 minutes, cooled and peeled
- 300g freshly picked wild garlic - washed
- 100g cornichons
- 50g capers
- 100g breadcrumbs
- 300g rapeseed oil
- 200g baby spinach
- Salt, pepper & sherry vinegar

METHOD

Blend all together until smooth. Taste and season with salt, pepper & sherry vinegar.

TRUFFLED PARMESAN CUSTARD

INGREDIENTS

- 200g Parmesan
- 6 whole eggs
- 50g double cream
- 1 tbsp truffle paste

METHOD

If using a Thermomix, place all into the jug except truffle paste, cook at 70°c, speed 4, for 7-8 mins until thickened. Mix through the truffle paste. Pour into a bowl and cover tightly with cling film and place in the fridge until cold.

If making without Thermomix, place all ingredients in a metal bowl and place over a pan of simmering water. Continuously whisk while cooking until the mix becomes thick and reaches 65-70°C. Chill as above.

FOCACCIA

INGREDIENTS

WILD GARLIC SALSA FOCACCIA
500g strong white bread flour
10g sugar
75g fresh yeast
680ml tepid water

METHOD
Mix the yeast and sugar into the water until it is fully dissolved. Then mix into the flour to form a light batter. Cover with cling film or a damp cloth and leave in a warm area for approx. 30-45 minutes until double in size, light and foamy.

Bread dough...
- 500g strong white bread flour
- 10g salt
- 30ml olive oil
- 4 tbsp of Wild garlic salsa verde

METHOD

Combine all the ingredients with the proved starter until fully incorporated and a dough forms. Gently fold in the wild garlic salsa verde. Transfer the dough to a bowl, cover with cling film or a damp cloth, and let it rest for 30–40 minutes, or until it has doubled in size. Once risen, carefully move the dough into a deep baking dish lined with greaseproof paper and greased with olive oil. Cover again and allow the dough to rise until it doubles in size. Drizzle with olive oil and sprinkle with sea salt. Bake for 10 minutes at 220°C, then reduce the temperature to 190°C and bake for an additional 15 minutes. Once done, remove from the oven and transfer to a cooling rack to cool to room temperature.

To finish
Local charcuterie such as "Salt Pig" four peppercorn coppa, "Salt Pig" air-dried smoked belly, "Salt Pig" Spiced loin.

To plate
Place some wild garlic salsa verde on the plate and place the hot or cold asparagus on top. Spoon some truffled parmesan custard onto the plate. Add some of the cured meats and finish with some wild garlic oil. Serve the warm bread on the side.

ASPARAGUS

Two spears of asparagus per person, ideally Wye Valley or another English variety.

Snap off the woody end of each spear. Blanch in salted boiling water for two minutes and then refresh in ice water to slow the cooking and keep the asparagus nice and green.

If serving cold, dress the asparagus in sea salt, olive oil & lemon juice. If serving hot, reheat on a grill or barbeque, and when nicely charred, dress as above.

HONEY, ORANGE AND PISTACHIO CAKE

BY SOPHIE PAGE

INGREDIENTS

CAKE

- 350g baking spread (Stork or Tesco work well)
- 2 tsp baking powder
- 350g caster sugar
- 7 medium eggs
- 350g plain flour
- 25g pistachios, blitzed to form a coarse dust
- 2 oranges

FILLINGS:

- 110g margarine or butter
- 380g icing sugar
- 50g white chocolate
- dash of milk
- 40g pistachios, blitzed
- Honey,
- Orange,
- Pistachio

METHOD

1. Lightly grease and line 4, 6" cake tins with parchment paper.
2. Preheat oven to 170°C.
3. Cream the fat and sugar on high speed until smooth.
4. Scrape down halfway.
5. Meanwhile, combine the flour, baking powder and nuts in a jug.
6. Crack eggs into a separate jug.
7. Turn the mixer to low and add the flour mixture and the eggs until thoroughly combined.
8. Zest the oranges and fold them in. Distribute the mixture between the four tins and bake for approximately 28 minutes.
9. Allow the cakes to cool in their tins before turning onto wire racks.
10. Place margarine into a mixing bowl and fit the balloon whisk.
11. Break down the fat for a few moments.
12. Break up the chocolate and place it in a bowl with the milk in the microwave for 1 minute on full power. Stir to combine.
13. Add the icing sugar to the margarine, mix slowly until combined, then add the chocolate and whip until smooth and creamy.
14. Level off the tops of the cakes & squeeze orange juice all over.
15. Spread 1/3 of the buttercream, sprinkle with nuts and drizzle with honey.
16. Repeat until fully stacked.

60

HAZELNUT PRALINE BON BONS

BY CHEF MICHAEL SCOTT

INGREDIENTS

- White Cocoa Butter
- 10ml Gold Cocoa Butter
- 20ml Blue Cocoa Butter
- 20ml Dark Chocolate
- 250g Double Cream
- 100g Isomalt Sugar
- 15g Dark Chocolate
- 125g Unsalted Butter, cut into cubes
- 20g Toasted Hazelnuts, chopped 50g

METHOD

1. Melt the Cocoa butter to 40-45°C.
2. Buff your polycarbonate mould with cotton wool pads.
3. Please make sure to paint whatever pattern you wish using your colours, but remember to allow each layer to set each time before adding more colour.
4. Melt 75% of the Dark Chocolate to 40-45°C, then stir in the remaining 25% until thoroughly melted.
5. Allow this to come down to 28°C.
6. Finally, bring the temperature back to 31°C, and you're ready to fill your cavities.

You can fill each cavity before tapping the mould to release any air bubbles, then turn the mould upside down over baking parchment and tap until the excess chocolate has all dropped out.

You can use a scraper to remove any excess chocolate on the mould.

METHOD - 2

For the filling:

1. Heat the Cream and Isomalt in a pan until the Isomalt has dissolved and come to a boil.
2. Pour over the Dark Chocolate and butter, and stir until thoroughly combined.
3. Finally, mix in the hazelnuts and allow them to cool to 28°C. Once cooled, transfer to a piping bag and fill each cavity, leaving a 2mm gap at the top.
4. Return to the fridge for at least 90 minutes to set firmly.
5. Once set, re-temper your chocolate as you did at the beginning, remove your mould from the fridge and allow it to warm slightly to room temperature.
6. Top off each cavity with Chocolate and scrape across the mould, leaving a clean mould and each cavity smoothly capped with Chocolate.
7. Return to the fridge for about 30 minutes to set fully; then, you're ready to release them.
8. Place two edges of the mould on a work surface and flip the mould upside down, slamming it flat onto the surface.
9. When you lift the mould your Bon Bons should release.
10. If some don't just give the mould a few light taps to release any tricky ones.

COCKTAILS

BY DAVID FROM WARWICKSHIRE GIN CO.

ELDERFLOWER FIZZ

PASSION FRUIT MARTINI/PORNSTAR

ELDERFLOWER FIZZ RECIPE

- Fill a shaker with 4 to 5 ice cubes.
- Add a double shot of Philosopher's Daughter Gin over the ice.
- Pour in 1 shot of lemon juice.
- Add a double shot of elderflower cordial.
- Shake well.

After pouring, top with prosecco for a delightful fizz. Garnish with dehydrated strawberries for an elegant touch.

PASSION FRUIT MARTINI/PORNSTAR RECIPE

- 4 – 5 cubes of Martini ice in shaker
- Double shot of Kingmaker Vodka over ice
- Double shot of passion fruit juice
- Single shot of Passeo passion fruit liqueur
- Shot of lime juice
- Drizzle of vanilla syrup

Instructions:
Shake all the ingredients well. Pour into a glass and garnish with dehydrated grapefruit.
Enjoy responsibly!

ESPRESSO MARTINI

BY STEVE FROM THAT GIN COMPANY

ESPRESSO MARTINI RECIPE

INGREDIENTS

- 25ml vanilla vodka,
- 25ml Tia Maria,
- 10ml sugar syrup (can be made at home with two parts sugar to one part boiling water) and then one fresh espresso shot.

METHOD

STEP 1
Begin by creating the sugar syrup. In a small pan, combine the caster sugar with the water over medium heat. Stir the mixture and bring it to a boil.

STEP 2
Once boiling, turn off the heat and let the mixture cool. Meanwhile, place two martini glasses in the refrigerator to chill.

STEP 3
After the sugar syrup has cooled, add one tablespoon to a cocktail shaker and a handful of ice. Add the vodka, Tia Maria, Espresso, and Coffee Liqueur.

STEP 4
Shake the cocktail shaker until its exterior feels icy cold.

STEP 5
Pour the mixture into the pre-chilled glasses.

STEP 6
Garnish with coffee beans.

PASSION FRUIT MARTINI

BY THAT GIN COMPANY

THAT GIN CO. PASSION FRUIT MARTINI RECIPE

INGREDIENTS

- 25ml vanilla vodka
- 25ml Passoa
- 50ml pineapple juice
- 25ml passion fruit puree
- 10ml sugar syrup.

The recipes have been suggested so people can easily get the ingredients from supermarkets.

METHOD

Combine Absolut Vanilla Vodka, Passoa, passion fruit puree, pineapple juice, and sugar syrup in a cocktail shaker.

Add a handful of ice and shake thoroughly. Strain the mixture into martini glasses, then garnish with half a passion fruit. Serve immediately alongside a shot of prosecco.

NEOPOLITAN PIZZA

BY CHEF MARIUS FROM THE LEOPARD SPOTS

INGREDIENTS

- 500g tipo 00 flour or the more readily available "strong white flour" will do
- 350ml cold water, preferably ice water, will help create a less sticky dough
- 2 tsp of salt
- 5g dried active yeast (not instant or easy bake)

METHOD

Step 1:
Dissolve the yeast. Dissolve the dried active yeast in 3 tbsp of warm water and stir until it's dissolved.

Step 2:
Mix the ingredients. Add all of the ingredients to a mixing bowl and stir with a large spoon until roughly incorporated. Leave the mixture for at least 20 minutes for the flour to absorb the water.

Step 3:
Knead the dough. Remove the mixture from the bowl and knead the dough until smooth. This will take at least 5 minutes.

For even more smoothness, let the dough rest under a warm, damp tea towel for another 10 minutes before forming the dough balls.

Step 4:
Making the dough balls. For 12-inch pizzas, cut the dough into 230g portions and use the method shown in the video to create three dough balls. For 10" pizzas, cut the dough into 170 g portions and use the same process to create 4 to 5 dough balls.

Step 5:
Proofing the dough. Place the dough balls at least 2cm apart in an airtight container and leave them to rise. This will take roughly 12 hours at room temperature or 24 in the fridge.

Leave at room temperature for a further couple of hours before stretching into bases.

METHOD CONT'D

The fridge method is more consistent because the temperature is constant instead of room temperature. Yeast works much slower in the fridge, requiring a longer proofing time.

With both methods, times may vary due to atmospheric conditions. The dough should be roughly double in size, full of air and spongy to touch.

Step 6:
- Stretch the dough. Lift a dough ball out of the container and place it into a bowl of flour.
- Turn the dough ball over and flour on all sides.
- Place the dough balls "skin side down" and use your fingertips to push the centre part of the dough balls so that you are left with a spongy circle of dough running around the edge.
- Lift the dough by the edge with the ends of your fingers and rotate like a steering wheel until the weight of the dough starts to stretch the base into a larger shape.
- Then, place the semi-stretched base on the back of your hands and pull your hands outwards while rotating until they are the required size.
- Finally, place the base "skin side down" onto a floured surface and top with your favourite ingredients. Don't take too long with the toppings, as the dough will stick to the work surface.

Final step:
Cook the pizza. If you have a pizza oven, cook it between 350°C and 500°C for 1 to 3 minutes, turning it halfway.

Use your eyes to determine when the pizza needs turning or has finished cooking.

Preheat the oven to its highest temperature for a conventional oven, place it on a preheated pizza stone or oven tray using a pizza peel, and cook until the ingredients have melted and the dough has browned.

LET'S EAT

This recipe is Neopolitan but you can be as creative as you like on the topping.

DUCK HAGU WITH FETTUCCINE

BY CHEF MARTIN
FROM TAVÓLA

FRESH TAGLIATELLE PASTA

INGREDIENTS

- 4 eggs
- 400g of flour (ideally 00 flour or all-purpose flour)

(Optional: You can purchase ready-made Grano Duro pasta) or 300g of tagliatelle from the store; follow the cooking instructions on the package

DUCK HAGU

INGREDIENTS

- 2 duck breasts (400g)
- 1 white onion
- 1 carrot
- 2 celery sticks
- 20g butter
- 1 spoonful of tomato concentrate
- 1/3 glass of red wine
- 5g fried rosemary
- 5g fried sage
- Salt
- Black pepper

FRESH PASTA

METHOD

1. Make a mound with the flour on a clean work surface and create a well in the centre.
2. Crack the eggs into the well, and using a fork, start beating the eggs, slowly incorporating the flour from the edges of the well.
3. Once the dough starts coming together, begin kneading it with your hands. Knead for about 10 minutes until the dough becomes smooth and elastic.
4. Shape the dough into a ball, wrap it in plastic wrap, and let it rest at room temperature for 30 minutes. This resting period allows the gluten to relax, making the dough easier to roll out.
5. After the dough has rested, roll it thinly using a rolling pin or pasta machine, and cut it into tagliatelle (long, flat noodles) or your preferred pasta shape.
6. Cook in boiling water for 4 minutes (tip - cook just after the duck is ready)

DUCK HAGU

METHOD

1. Start by chopping the carrot, celery, and onion to make the base for the soffritto.
2. The pieces should be small and uniform, about 2-3 millimetres on each side.
3. Heat a pan with a tablespoon of butter and a sprinkle of pepper; when the pepper starts to sizzle, it's time to add the duck skin.
4. The duck skin flavours the sauce but will not be served, so it should be removed once the cooking is complete.
5. I leave it whole to make it easier to remove; after turning it over during cooking, it will be easier to remove.
6. Once it starts to brown and "melt," add the vegetables and make a classic soffritto.
7. As soon as the onion becomes golden, add the diced duck meat and let it brown over low heat so it gains colour without burning.
8. If you notice the soffritto starting to dry out too much, add a little salt to the meat (this will make it release some juices) or simply some of the pasta water that is boiling on the stove.
9. Before adding the tomato paste, remove the duck skin from the pan and add some fried rosemary.
10. After that, add a tablespoon of tomato paste and, after mixing everything and the pan has reached its maximum heat, add a quarter glass of red wine.
11. Cook for another 15 minutes to evaporate the alcoholic part.
12. Now, the duck ragù is ready; keep it slightly soupy so you can toss the pasta to absorb the taste.
13. Once the fresh pasta is made and cooked, add it to the duck ragù to coat it thoroughly.
14. For garnish, fry a few leaves of sage until crispy and place them on top of the dish for an aromatic finishing touch. You can also add freshly shaved black truffle, though this is optional based on your preference.

OCTOPUS LASAGNE

BY CHEF ANGELO - RODIZIO RICO

OCTOPUS LASAGNE

INGREDIENTS

- 1kg octopus (chopped)
- 4 tbsp extra virgin olive oil,
- 1 tsp chilli flakes
- 1 tsp black pepper,
- 2 sticks celery,
- 1 large grated carrot,
- 1 big potato (5/6 baby)
- 1 chopped red onion,
- 3 shallots
- 3 cloves garlic,
- 1 glass red wine,
- 2 tbsp tomato paste,
- 300g plum tomatoes peeled,
- 1 handful chopped parsley,
- 1 handful of chopped coriander
- 500g lasagne sheets, no need to boil.
- 100g grated grana cheese,
- 150g cheddar cheese,
- 1/2 red, green and yellow peppers
- 1 red onion,
- 5 cherry tomatoes,
- 1 tsp oregano,
- 1 tsp smoked salt

WHITE SAUCE

INGREDIENTS

- 750ml milk,
- 2 tbsp butter,
- 2 tbsp (flour / cornstarch)
- 1 tsp nutmeg,
- 1 tsp salt,
- 1 tsp black pepper,
- 1 tbsp garlic powder

METHOD

1. In a saucepan, melt the butter on medium heat.
2. Add the garlic, black pepper, and nutmeg, then add the milk, salt, and flour.
3. Mix thoroughly on medium heat until it reaches a creamy consistency.

ASSEMBLING THE LASAGNE

- Spread a layer of white sauce in the cooking dish, then place lasagne sheets on top.
- Add a layer of octopus ragù, followed by another layer of lasagne sheets.
- Repeat with four layers of Octopus ragù and five layers of lasagne sheets.
- Top with white sauce and cheese.
- Garnish with chopped peppers, onions, and oregano.
- Bake at 180°C for 25-30 minutes or until the lasagne sheets are cooked.

Enjoy!

Chef Angelo's tip: Add 200g of the white sauce to the octopus ragu. That will make the Ragu creamier.

86

METHOD

1. Add olive oil, red onions, shallots, garlic, celery, carrots, half of the parsley, and chilli flakes in a saucepan. Cook for 2 minutes.
2. Add the octopus and cook for 1 minute.
3. Add the wine and let it cook for 2 minutes until the alcohol evaporates.
4. Add the tomato paste and the plum tomatoes.
5. Add salt and pepper.
6. Cook on low heat for 2 hours.
7. Remember to add the potato mix and stir thoroughly so it doesn't stick to the bottom of the pan.
8. When ready, reserve aside.

PANCETTA AND MUSHROOM RISOTTO

BY CHEF ANGELO - RODIZIO RICO

INGREDIENTS

- 3 tbsp extra virgin olive oil,
- 3 shallots,
- 1 tsp chilli flakes ,
- 3 cloves garlic,
- 200g pancetta,
- 1 tsp black pepper
- 300g carnaroli rice
- 300ml prosecco,
- 100g porcini mushrooms
- 3 king oyster mushrooms,
- 4 shitake mushrooms
- 30g fresh parsley
- 1 lemon zest
- 5 tbsp butter
- 100g parmigiano cheese
- 30g pecorino cheese
- 6 baby burratas
- Garnish with black truffle
- edible flowers, micro veg

METHOD

1. Start with the broth used to cook the rice (two litres of water, one carrot, one celery stick, one onion, dash of salt, and one chicken stock cube).
2. In a pan, add the olive oil and butter. Add the shallots, garlic, pancetta, black pepper, chilli flakes and fry for two minutes.
3. Add some of the porcini mushrooms, carnaroli rice, and prosecco.
4. Once the alcohol evaporates, add some broth and all the mushrooms (porcini, king oyster, shiitakes and chestnut).
5. Mix thoroughly to prevent the rice from sticking to the bottom of the pan, and keep adding broth slowly until the rice is cooked.
6. When the rice is cooked, turn the heat off, add the rest of the butter, and add the cheese. Give it a good mix, put the lid on the pan, and let it rest for two minutes. That will help the ingredients combine, making a nice creamy risotto.
7. Garnish with black truffle, olive oil, burrata, edible flowers, micro veg, and black pepper.

Enjoy!

BIA'S KITCHEN
Show

91

BIA'S TWIST ON HER FAVOURITE RECIPES

BIA'S FEIJOADA BLACK BEAN STEW

FEIJOADA - BLACK BEAN STEW

INGREDIENTS

Serves 6-10 portions

- 250g dried black bean, soaked overnight, then drained
- 100g streaky smoked bacon, cut into slices
- 500g pork rib
- 3 chorizo cooking sausages
- 500g pork shoulder, cut into 5cm cubes
- 300g top rump steak cut into cubes
- 3 onions, chopped
- 6 garlic cloves, finely chopped pinch of chilli flakes 1tbsp dry coriander
- 1 tbsp smoked paprika
- 1 tsp cumin powder
- 1 tsp black pepper powder
- 1 tbsp oregano
- 6 bay leaves
- 2 tbsp white wine vinegar
- 1 handful of coriander

ACCOMPANIMENTS

Rice

- 370g basmati rice
- 5 tbsp coconut oil or any oil you like
- 3 cloves of chopped garlic
- 30g salt

To serve
Steamed rice, chopped parsley or coriander, hot pepper sauce and wedges of oranges

Feijoada looks like lots to do, but it is straightforward - You need to put the beans and the meat in a big pot or pressure cooker, cover them with water and cook for around 1 hour or until the meat is tender.

STEP 1
Heat a large, heavy-based saucepan with a fitted lid. Add the bacon and fry until crisp. Remove and keep the oil in the pan. Sear the ribs, sausages, beef, and pork shoulder in batches. Season each batch with salt and pepper.

STEP 2
Add the onion, garlic, chilli, and fry for 8 minutes or until soft. Add dry coriander, black pepper, cumin, paprika, oregano, bay leaves, and the beans.

STEP 3
Cover with just enough water to cover, about 650ml. Bring to a boil and reduce the heat to a low simmer. Cover and cook for 2 hrs until the beans are soft and the meat is tender. You can also use a slow cooker using the short method (4 hr) or make a quick version using a pressure cooker in batches for 30 mins each. Another method is to cook it in the oven for 3-4 hrs at 160°C/ 140°C fan/gas 3.

STEP 4
Serve with rice, spring greens saute in butter, fresh coriander, hot pepper sauce and orange slices.

BIA'S BRAZILIAN SEAFOOD CURRY

INGREDIENTS

- 350g langoustine
- 400g prawns
- 300g mixed fish fillet
- 5 chopped lime leaves
- 1 red pepper
- 1 green pepper
- 1 yellow pepper
- 2 large onions
- 2 fresh chillis chopped
- 1 tbsp paprika
- 1 tbsp dry coriander powder
- 6 tbsp fish sauce
- 4 cloves of crushed garlic
- 30g of chopped ginger
- 350ml coconut cream
- 350ml double cream
- 600g chopped tomatoes
- 30ml of Dendê oil (palm oil)
- 2 handfuls of fresh coriander

METHOD

1. Add Dendê oil, onions, and garlic to a large pan, and cook until the onions are slightly softer.
2. Add ginger, peppers, paprika, dry coriander, chilli, and fish sauce. Mix well, then add the langoustine, lime leaves, and desiccated coconut.
3. Mix the double cream with the coconut milk and pour into the pan, then add the chopped tomatoes, stir a bit and add the rest of the fish and prawns.
4. Cover with the lid and cook for 15 minutes or until the fish is cooked.
5. Finish with a handful of fresh coriander, lemon juice and a drizzle of Dendê oil.
6. Enjoy with white rice.

CHILLI CON CARNE

INGREDIENTS

- 500g mince beef,
- 1 large chopped onion
- 1 minimum red and green pepper,
- 6 chopped tomatoes,
- 1 tbsp dark chocolate,
- 5 cloves of crush garlic,
- 400g red kidney beans,
- 1 tsp gluten free flour,

- 1 tbsp chilli powder,
- 1 tbsp cumin powder,
- 1 tbsp oregano
- 1 tbsp smoke paprika
- 1 tsp cayenne pepper,
- 1 tbsp brown sugar or honey,
- salt as you like, fresh coriander and lime juice.

You can use tin of chopped tomatoes if you like

METHOD

1. Brown the minced beef for 5 minutes with onions and garlic.
2. Add the chocolate, sugar, tomatoes, spices, kidney beans, and 200ml of water, and bring to a boil.
3. Cover and simmer gently for 15-20 minutes; don't forget to stir occasionally.
4. Last but essential is fresh coriander, juice, and zest of two limes.
5. Enjoy with tacos or rice.

BEEF STROGANOFF

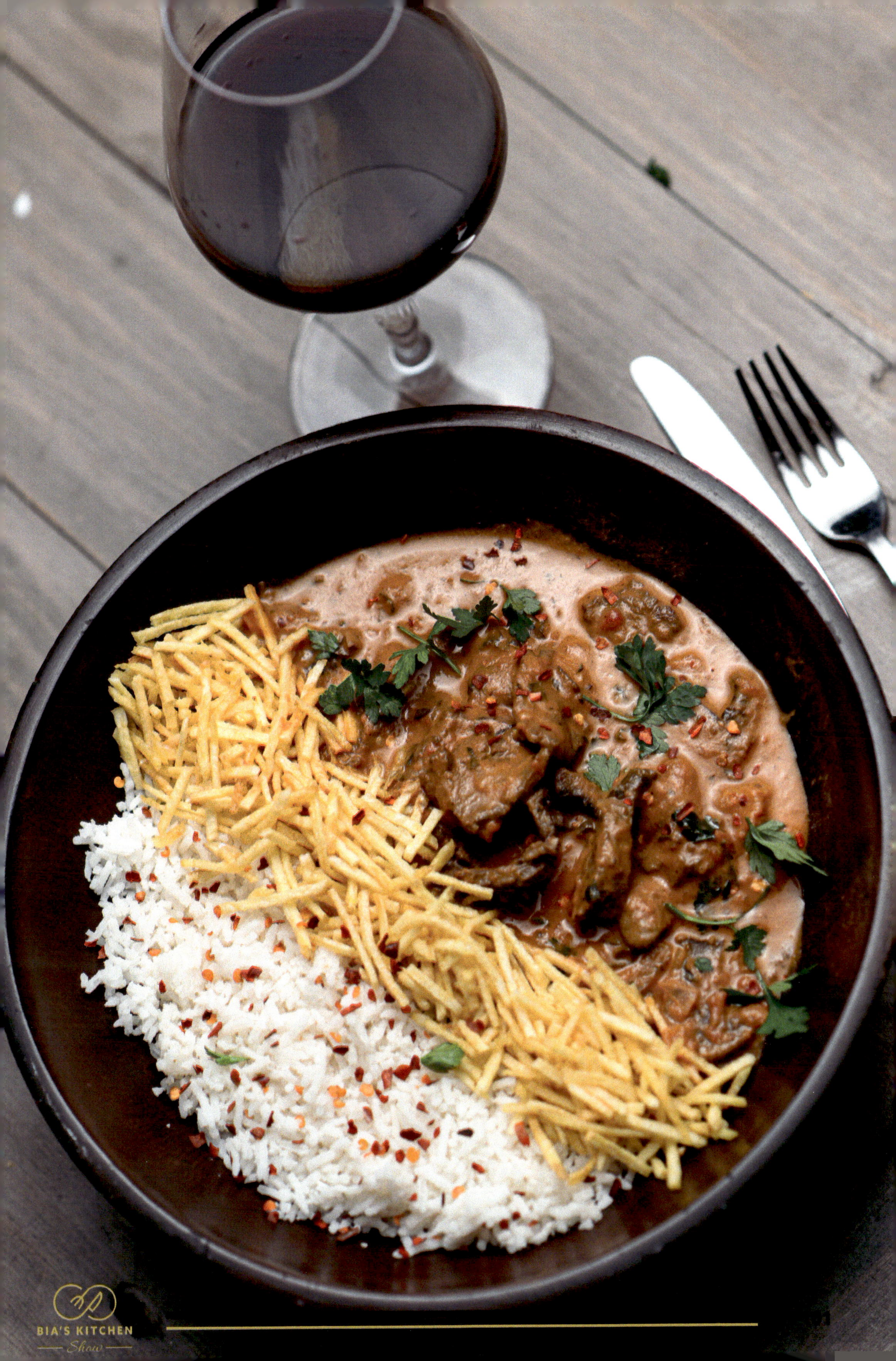
BIA'S KITCHEN
Show

INGREDIENTS

- 500g Beef skirt
- 250g Mixed mushrooms
- 1 tbsp dry coriander powder
- 1 tbsp paprika
- 1 tsp black pepper
- 1 Large glass of red wine
- 4 cloves of crushed garlic
- 2 tbsp olive oil
- 4 tbsp Worcestershire sauce
- 1 tbsp pink salt
- 350g double cream
- 600g chopped tomatoes
- 1 tbsp sugar
- 1 tsp mustard
- 1 Large chopped onion
- 2 Handfuls of fresh parsley

METHOD

1. Cut the beef into small slices, add olive oil, garlic, dry coriander, Worcestershire sauce, salt, black pepper (optional)
2. Mix and cover to marinate overnight in the fridge or two hours before cooking.
3. Add the marinated steak to a pan; oil is unnecessary.
4. Grill until it is cooked. Separate the beef and put it aside.
5. Add chopped onions and mushrooms in the same pan you used to grill the beef.
6. Cook until it is soft and tender.
7. Add the meat and the wine and mix; add smoked paprika, sugar, a glass of wine, chopped tomatoes and mustard, and bring to a boil.
8. Cook for 5 minutes, add double cream, and cook until it is creamy.
9. Before serving, add the chopped parsley, and for an extra kick, sprinkle some chilli flakes.
10. Serve with rice or pasta.

BIA'S FUSION CHICKEN CURRY

103

INGREDIENTS

- 500g boneless chicken thighs
- 1 large chopped onion
- 6 cloves of crushed garlic
- 30g ground cumin
- 30g dry ground coriander
- 20g smoked paprika
- 1 tbsp of fresh chopped ginger
- 30g mixed curry spices
- 350ml double cream - can be lactose-free
- 350ml coconut cream milk
- 1 tbsp dark brown sugar
- 30g fresh curry leaves - can be dry curry leaves too.
- 100g desiccated coconut
- 100g chopped fresh coriander
- 1 tsp salt
- Zest and juice of 2 limes.
- Fresh chilli (optional)

METHOD

1. Marinate the chicken overnight or for 1 hour before cooking with lemon juice, salt, cumin, garlic, dry coriander, and paprika.
2. In a hot pan, add the chicken.
3. After that, marinate it and fry until golden.
4. Pour in the ginger, mixed curry spices, onions and curry leaves, and cook for 5 minutes with the lid off.
5. Add the coconut milk, double cream and desiccated coconut and cook for 10 minutes, until creamy.
6. Before serving, add fresh coriander, lime zest and juice.
7. Enjoy with nice garlic rice and naan bread.

GREEN PEA SOUP WITH CRISPY BACON

INGREDIENTS

- 500g green split peas
- 1 large chopped onion
- 5 cloves crushed garlic
- 1 tbsp dry coriander
- 1 tbsp finely chopped fresh ginger
- 2 vegetable stock cubes
- 600g chopped smoked bacon
- 300 ml double cream

METHOD

1. Heat the bacon in a saucepan over a medium heat.
2. Cook until it is light brown and crunchy. Set it aside.
3. In the bacon fat, add the onion, garlic, and ginger and fry for 3-4 minutes, until softened.
4. Add the dry peas and vegetable stock and mix in 1.5L of hot water. Add dry coriander and bring to a boil.
5. Reduce the heat and simmer for 50 minutes or until the peas are soft. Add extra water if necessary. (You can also cook in the pressure cooker for 30 minutes or slow cooker for 2 hours.)
6. Add the cream and use a hand blender to liquidise the soup. Season to taste and serve in a warm bowl garnished with crunchy bacon and fresh coriander.
7. Enjoy with lovely fresh bread and butter.

BRIGADEIRO - BRAZILIAN CHOCOLATE TRUFFLES

INGREDIENTS

Bia's Classic Brigadeiro Brazilian Truffles
(Makes 8-10 servings)

- 50g butter
- 14 oz sweetened condensed milk (395 g)
- ¼ cup dark cocoa powder (30g)
- 1 cup chocolate or multi-coloured sprinkles (160g), as needed to decorate.

These traditional Brazilian sweets are the perfect dessert for any occasion!

These brigadeiros are rich, creamy, and delicious made with just four ingredients.

A great way to get the children involved and an exciting alternative for Easter.

METHOD

1. In a pot over low heat, melt the butter, condensed milk, and cocoa powder, stirring continuously until you can see the bottom of the pot for 2-3 seconds when dragging a spatula through.
2. Pour onto a greased plate, then chill for 1 hour.
3. Shape and roll the chilled mixture into balls.
4. Roll the balls in chocolate sprinkles.
5. Enjoy!

PASSION FRUIT MOUSSE AND DARK CHOCOLATE CHEESECAKE

INGREDIENTS

Mousse
- 300ml pure passion fruit pulp
- 300g double cream
- 300g condensed milk

Topping
- 4 fresh passion fruit
- ¼ cup sugar – for the sauce on top.
- 50ml water

Cheese Cake
- 200g/7oz dark chocolate, minimum 64% cocoa solids, finely chopped
- 10g unsalted butter, at room temperature
- 300ml/10fl oz double cream
- 2 tsp caster sugar
- 200g cream cheese

METHOD

1. Blend the passion fruit pulp, condensed milk, and double cream for 5 minutes in the liquidiser.
2. Pour the mousse into a large serving bowl or individual serving glasses. Refrigerate for at least 3 hours.

For the topping
1. Combine the fresh passion fruit pulp (with seeds) and the sugar in a saucepan. Bring it to a boil, then lower the heat and cook briefly until it thickens slightly.
2. Let it cool to room temperature. You can then serve or refrigerate until ready to use. Top the mousse with the passion fruit sauce and serve!

Cheese cake
1. Tip the chocolate and butter into a bowl and set aside.
2. Pour the cream into a saucepan, add the sugar and bring to the boil, stirring occasionally. Remove from the heat, leave to cool for 30 seconds and then pour over the chocolate mixture. Leave the hot cream to melt the chocolate without stirring for 1 minute and then stir in one direction only, using a wooden spoon or rubber spatula, until smooth and glossy.
3. Once the ganache is smooth, leave it to cool and thicken slightly before using.
4. Then add the cream cheese and mix slowly until it's all combined together.

BIA'S CHOCOLATE & PISTACHIO TOWER CAKE

INGREDIENTS

Bia's Chocolate & Pistachio Tower Cake
(Serves 5-6)

- 225g self-raising gluten-free flour
- 300g sugar
- 114ml milk
- 114ml oil
- 114g butter
- 220ml water
- 1/2 tbsp baking powder 2 eggs
- 4 tbsp dark chocolate
- 1 tbsp Nescafé coffee granules

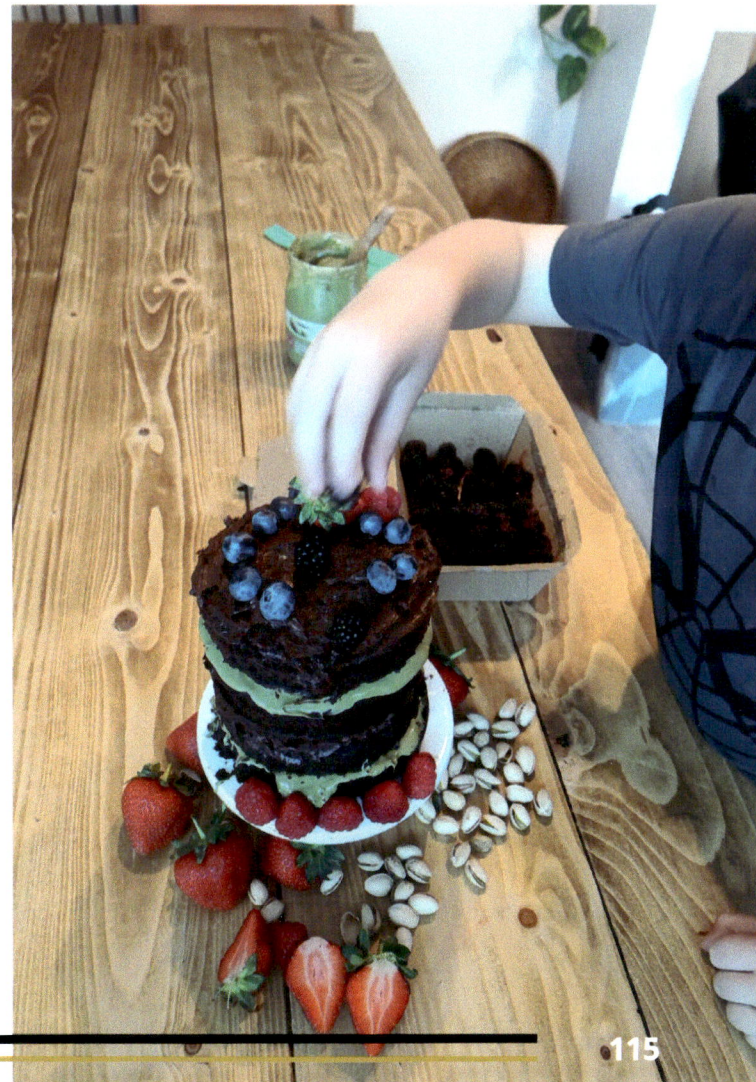

METHOD FOR THE CAKE

Step 1
Heat oil, butter, dark chocolate powder, water, and coffee in a pan until boiling.

Step 2
Stir in flour, sugar, then mix well and add eggs combined with milk.

Step 3
Add baking powder, mix again, and pour the mixture into an 8-inch baking tray.

Bake for 45 minutes at 170°C oven – leave to cool before assembling the cake.

For the filling:
Use 300g of ready-to-spread Pistachio cream bought online.

Here's a link to Pisti Sicilian Pistachio Cream Spread - 600g:

https://amzn.eu/d/1aHjll9

METHOD

For the chocolate ganache:

1. Combine 200g/7oz finely chopped dark chocolate with 10g of room-temperature unsalted butter.
2. Heat 300ml double cream with 2 tsp caster sugar until boiling.
3. Pour the hot cream over the chocolate mixture, let it sit for a minute, then stir to a smooth consistency.

Assembling the cake:

1. Cut the cake into three layers and add the filling between each layer.
2. For added flair, consider incorporating fruits.
3. This cake is sure to be a hit with the whole family!

KIBE

INGREDIENTS

- 1 kilo fine Bulgar wheat,
- 1 kilo mince meat
- 2 onions,
- 1.5 tbsp Kibbeh Spices
- 1.5 tbsp cornflour
- 1 tsp sevenspice
- ½ tsp black pepper
- 1 tbsp salt
- 2 tbsp olive oil
- 4 cloves of garlic
- handful of fresh mint chopped fine

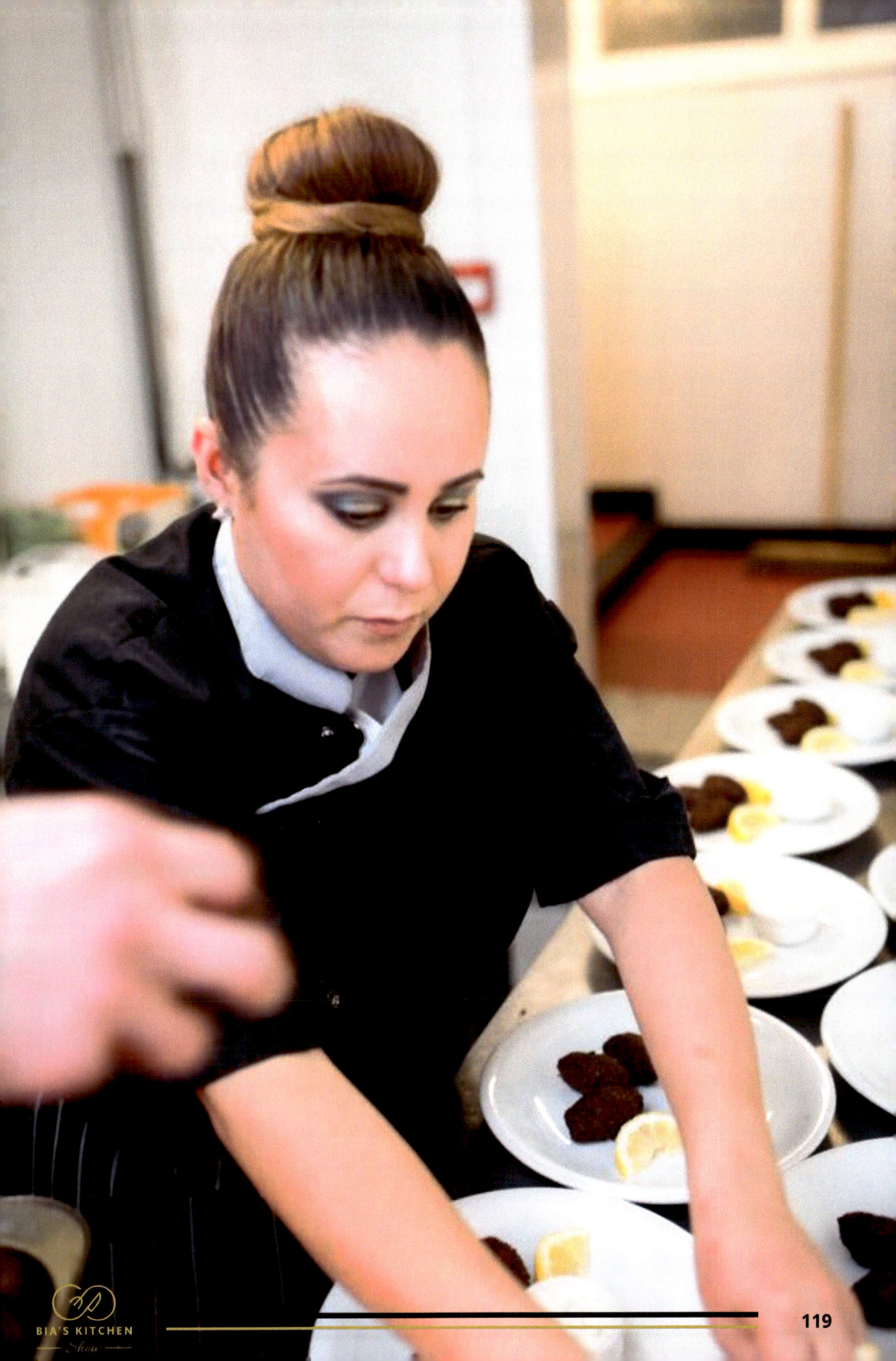

METHOD

1. Soak the Bulgar wheat in enough water to cover and set aside.
2. Drain the Bulgar wheat and add to a large bowl.
3. Take out the kilo of meat and prepare a food processor, or use your hands to mix it all.
4. Add the 2 onions to the food processor along with the seven spices, crushed garlic, black pepper, and salt.
5. Mix well. When you have combined all the meat and Bulgar wheat in a bowl, add the blitzed onion spice mix, then bring everything together, preferably with your hands.
6. If it is too dry, add a bit of water. You don't want it too loose, sticky, or crumbly, so it should hold together nicely.
7. If you don't mind tasting some raw meat, check for saltiness and adjust as necessary.
8. Add approximately half a cup of olive oil to a small bowl. Dip your fingers into the oil before shaping the kibbeh to avoid sticking.
9. Clean your table and arrange a workflow with a large, clean tray. Take some casing and roll it into golf ball size; compress it well with your hands and shape it into classic kibbeh morsels.
10. Line up the kibbeh on the tray until you finish the mixer.
11. Deep fry your kibbeh morsels in sunflower or vegetable oil and serve with yoghurt dip.

PÃO DE QUEIJO - CHEESE BREAD

INGREDIENTS

(Makes approx. 24 balls)

- 500g Manioc starch
- 100g parmesan grated cheese
- 200g grated mozzarella
- 100g grated cheddar cheese
- 2 eggs
- 150ml milk
- 150ml water
- 100ml coconut oil (or any oil you like)
- 1 tsp pink salt

METHOD

1. Add the milk, water, oil, and salt to a pan and bring it to a boil.
2. In a large bowl, add the Manioc starch, pour in the hot liquid, and mix it well. Let it cool down, then add the eggs, parmesan, and cheddar.
3. Mix well, then add the mozzarella.
4. Mix it all together until soft and starts to come away from the sides of the bowl.
5. Line a baking sheet with greaseproof paper. With wet hands, divide the mixture into Twenty-four pieces and roll them into shiny balls, lining them up on the tray as you go.
6. Pre-heat the oven to 180°C/gas mark 6.
7. Bake uncovered for 18 minutes or until golden.
8. Enjoy it warm with a glass of red wine or a black coffee.

TIP - These are fantastic cooked from frozen if you want to get ahead. Bake for 20 minutes at 180°C or until lightly golden. Children love to help to make them!

STEAK IN THE PAN

INGREDIENTS

- 300g beef skirt
- 30g butter
- 3 cloves garlic
- Hand full of fresh rosemary
- 1 tsp salt flakes

METHOD

1. Place the pan on the hob and let it get super hot!
2. Put the salt on the steak and then put it straight in the pan.
3. Grill for 2 minutes on each side for medium rare. Take out the steak, roll it in foil, and place it to one side for 2 minutes.
4. Add garlic butter and rosemary in the same pan you grilled the steak and bring the steak back in the pan to give that caramelised look and taste.
5. Serve rice, farofa and vinaigrette salsa. Or just with chips!

Tip: To know when your pan is ready to put your steak in, wait until it stops smoking; then, it will be ready. This recipe can be used for any steak you love: ribeye, rump steak, etc..

BRAZILIAN BBQ

INGREDIENTS

- 2 Ribeye steak or cap rump (PIcanha)
- 4 chicken drumsticks marinaded with juice of one lemon,
- 1 tbsp mayonnaise,
- 1 tsp salt,
- 2 cloves of crushed garlic linguica Calabresa
- Brazilian sausage 100g chicken heart marinaded in 1 orange juice and two cloves crushed garlic
- 2 tbsp rock salt for the steak

Vinagrette
- 1 tomato chopped
- 1/2 onion chpped
- 1/2 green pepper
- 12 tbsp olive oil
- 2 tbps apple vinegar

Farofa - yucca flour
- 1/2 onion chopped
- 3 cloves of crushed garlic
- 50g butter
- 1 pinch of salt

SIDES
Garlic Rice
- 3 cloves of crushed garlic
- 1 tbsp coconut oil or any oil you like

MEATS

METHOD

1. Build a wood or charcoal fire and preheat to medium-high (400 degrees).
2. Insert the skewers on the BBQ grill, sausages on the top row, and chicken steak on the bottom.
3. Spit-roast until each is cooked to taste: about 35 to 30 minutes for the chicken and 20 minutes for the sausages.
4. Rotate the steak every 3 minutes, depending on the heat of your fire.
5. Cook the chicken through (to an internal temperature of 165 degrees).
6. Serve the beef crusty on the outside and medium-rare (130 to 135 degrees) inside.

MOLHO A CAMPANHA - VINAIGRETTE SALSA

METHOD

Step 1
Place the wine vinegar, salt, olive oil, and crushed garlic in a nonreactive mixing bowl and whisk until the salt dissolves.

Step 2
Add the green pepper, tomato, onion, parsley or coriander and stir to mix. Taste for seasoning, adding more salt and/or vinegar as necessary

FAROFA - YUCCA FLOUR

METHOD

Step 1
Add the butter and onions to a frying pan, cooking until golden. Add the garlic, and cook until light brown.

Step 2
Add the yucca flour and keep mixing until the flour is crispy. Ready to serve

GARLIC RICE

METHOD

Step 1
Add the oil and garlic in a pan, and cook until lightly golden.

Step 2
Add the basmati rice; no need to be washed. Mix for 2 minutes.

Step 3
Add enough water to cover the rice by 2 centimetres, and let it cook on high heat until the water evaporates.

Step 4
Turn off the heat, cover the pan, and let it finish cooking on the hot pan.

BEEF CARPACCIO

INGREDIENTS

- 50g fillet steak
- 1 tbsp caper
- 2 tbsp extra virgin olive oil
- 10g parmeson
- salt and pepper to your taste
- Handful of rocket
- 2 tbsp balsamic vinegar or glase

Bia's tips: Garnish with cherry tomatoes or slices of strawberries.

METHOD

1. Freeze beef until very firm but not rock solid, about 1-2 hours.
2. While the meat is in the freezer, chill the plates in the fridge.
3. Once the beef is firm, slice thinly against the grain and place slices on chilled plates.
4. Top each serving with a handful of rocket, capers, tomatoes, and parmesan cheese.
5. Drizzle with vinegar or balsamic and olive oil, and add a pinch of freshly cracked black pepper and sea salt.
6. Serve immediately.
7. You can add a drizzle of honey for extra flavour.

DETOX OVEN OMELETTE

INGREDIENTS

- 6 eggs
- 100g chopped green beans
- 100g chopped asparagus
- 1/2 chopped red pepper
- 1/2 chopped yellow pepper
- 3 cloves of crushed garlic
- 30g baby corn
- 30g dried tomatoes
- 100g grated parmesan cheese
- A handful of baby spinach chopped
- 1/2 chopped onion
- 50ml coconut milk or any milk you like
- 1 tsp dry coriander
- 1 tsp oregano
- Salt and pepper to your taste
- A handful of fresh basil.

METHOD

1. Chop all vegetables and mix with the beaten eggs.
2. Add salt, dry coriander, oregano, and cheese to an 8-inch baking tray.
3. Cover and take to the oven at 180°C for 30-45 minutes or until it is cooked.
4. Serve with rice.

ENJOY WITH...
This dish is incredibly versatile. You can enhance it with shredded chicken, bacon, ham, or enjoy it as is. Pair it with rice, a green salad, or savour it on its own. Bon appétit!

BRAZILIAN SALT COD PIE - BACALHAU ESPIRITUAL

INGREDIENTS

- 600g Shredded bacalhau (salted cod)
- 1L milk
- 6 egg yolks
- 250ml double cream
- 6 tbsp flour
- 2 grated carrots
- 50g butter
- 4 tbsp olive oil
- 100g grated parmesan
- 1 tbsp dry coriander powder
- 2 grated onions
- 2 tbsp Worcestershire sauce
- 300g cream crackers crumbled
- fresh coriander to dress!

METHOD

1. Add water to the bacalhau in a saucepan and bring it to a boil. Keep the heat medium to low to keep it from boiling.
2. Melt the butter in another saucepan. Add half of the milk mixed with flour. Pour this mix into the saucepan, boil, and add the other half of the milk with the egg yolks.
3. Cook until the mixture leaves the bottom of the pan, but do not stop whisking. Set aside.
4. Add the crackers to a mixer and mix until they are crushed. Add parmesan to this mixture. Set aside.
5. After the Bacalhau is cooked, we add the shredded bacalhau, carrots, double cream, onions, dry coriander, Worcestershire sauce, olive oil, and fresh coriander and mix it all together.
6. Set everything in the tray and add the white sauce. Then, add the crackers and parmesan mixture.
7. The oven should be set for 35 minutes at 180°C or until golden brown.
8. Enjoy

CREAMY RISOTTO WITH ROAST HONEY BEETROOT

BEETROOT WITH HONEY AND CARAMELISED NUTS

INGREDIENTS

- 4 beetroots cut into squares
- 100g crushed razor nuts or almonds
- 1 cup of sugar
- 1 tbsp butter
- 2 tbsp olive oil
- Salt and pepper to your taste.

RISOTTO

METHOD

1. Melt the butter and sauté the sliced onions in a medium-sized pan until they become soft.
2. Incorporate the risotto rice, stirring for about a minute, then increase the heat and pour in the wine and mustard, stirring until the wine is fully absorbed.
3. Gradually add the hot stock, ladling in one portion at a time, allowing each ladleful to be absorbed before adding more.
4. Continue stirring and ladling until the rice reaches an al dente texture, which should take approximately 18 minutes. Then fold in the cheese, mixing it into the rice until it melts.
5. Remove the pan from the heat while continuing to stir, then serve the risotto in warmed dishes, garnishing with some chopped chives.

BEETROOT AND CARAMELISED NUTS

METHOD

1. Turn the oven on to 180°C.
2. Add the beetroot to roast in a tray with salt, pepper, and Olive oil.
3. Cook for 30 minutes or until it is tender.

Caramelised Nuts
1. Add sugar to a pan and let it cook until it is light brown.
2. Add the butter and the nuts and cook until the sauce is thick, leaving the pan.
3. If it goes hard, you can put a bit on the side. It is ready.
4. Pour the sugar sauce on a surface you can bang to break it into pieces.
5. There you have your crunchy nuts.
6. Pour the beetroot on top of your risotto, sprinkle the crunchy nuts, and enjoy it.

BIA'S CHRISTMAS TREE

This festive Christmas tree allows all the children to join in the fun of preparation and tasting!

The cheeses I selected include:

- Wensleydale with cranberries for the trunk
- Mature white cheddar and medium-coloured cheddar
- For the top and the star, I used apricot Wensleydale.
- The fruits featured are blackberries, green grapes, and red grapes.

To add a touch of dark Christmas green, I incorporated some fresh rosemary, and on the side of the board, I placed some Brazilian mini toasts.

Merry Christmas everyone!

About our guests

About our guests

ABOUT TROPICAL BRAZIL

EXPLORING THE FLAVOURS OF TROPICAL BRAZIL FOODS WITH CHEF TATIANA

Tropical Brazil Foods is led by Tatiana, a self-taught Brazilian chef who offers a variety of mouth-watering dishes, including the popular pastel, at her food truck in Coventry. Tatiana was our very first guest for the show, and we enjoyed cooking some of my favourite Brazilian street food, pastel de carne. It was a delightful experience that brought back fond memories of the bustling streets of Rio de Janeiro.

WHERE TO FIND TROPICAL BRAZIL FOODS

@tropicalbrazilfood - Instagram
WhatsApp +44 078 6806 9339

149

ABOUT NIKKI RYAN

Nikki's content combines her passions of food, shopping and motherhood.

Nikki says *"I'm all about healthy eating recipes, support, inspiration, beauty, fashion, IVF, pregnancy and now also being a mum to twins."*

HER STORY

After losing three stone in 3 months with Slimming World, Nikki created a popular social media platform sharing recipes, tips, and lifestyle content.

She has been featured in magazines, was a successful Team Developer at Slimming World, and is an award-winning Gold Consultant. Nikki's journey includes overcoming infertility to become a twin mum, which has added depth to her social media presence.

WHERE TO FIND NIKKI

@nikkiryan - on intagram
https://www.fopperholic.co.uk/
ACCESS NOW

150

INTRODUCING THE PAPRIKA CLUB RESTAURANT

PAPRIKA CLUB

The Paprika Club is a family run-business serving authentic Indian food in Leamington Spa for over 27 years.

WHAT THEY ARE ABOUT

Our aim is to provide a unique dining experience serving exquisite Indian cuisine representing our culture and legendary gastronomy with typical dishes using a creative and talented touch. We pride ourselves on quality and most importantly our customers.

AWARDS

Best Curry Restaurant in Britain at the Curry Life Awards, staged by Curry Magazine 2021

Paprika Club was named the West Midlands' Asian Restaurant of The Year 2024.

@paprikaclubrestaurant - on Instagram
www.paprikaclub.co.uk

EMI'S LITTLE BAKERY

Emi's Little Bakery is famous for its delightful brownies, cakes, and cookies. Recently, Emi expanded her offerings by introducing a new line of ice cream in collaboration with professional rugby player Zach Kibirige under the brand The Inside Scoop. Even if you haven't ordered from Emi's Little Bakery directly, you might have indulged in her tasty brownies, cookies etc.. at different cafes across Warwickshire, primarily available to buy at Aubrey Allen in Leamington Spa.

WHERE TO FIND EMI'S LITTLE BAKERY

Instagram - @emislittlebakery
www.emislittlebakery.shop

SCRATCH COOK SOCIAL

WHAT THEY ARE ABOUT

Cooking from scratch is at the heart of everything we do and with nature providing such a bounty of goodness, it is by marrying the produce with traditional cooking techniques that we aim to elevate our recipes to excite and delight all your senses.

RAJ'S COMMENT ON APPEARING ON THE SHOW

"I was extremely honoured to be invited to film a 60-minute episode with Bia, where I was able to share my journey 'from corporate exec to foodie entrepreneur' and all the important influences in my life that have informed the type of business I have created".

"It was lucky enough for my path to cross with that of the amazing Bianca Rodrigues Perry A mother, an entrepreneur, a restaurateur, a producer and moreover, a huge personality".

WHERE TO FIND SCRATCH COOK SOCIAL

Instagram - scratchcooksocial
www.scratchcooksocial.com

153

BANH MI CA PHE

Since 2014, its popularity has surged, introducing this Vietnamese favourite to a global audience. Originating as a humble street cart in late 2019, Banh Mi Ca Phe swiftly evolved into a beloved food brand across the West Midlands. Despite the global challenges, they established their brick-and-mortar store on Leamington Spa's Regent Street in April 2021, progressing successfully ever since.

BBC TALKS ABOUT BANH MI

A decade ago, the BBC recognised Banh Mi as one of the world's best sandwiches.

BIA ABOUT THAO, THE CHEF BEHIND BANH MI CA PHE

We had a pleasure to interview the co owner Thao where she shared her delicious recipe with our audience. We had fun on the set!

FIND OUT MORE

Instagram - @Banhmi_ca_phe
www.banhmicaphe.co.uk

THE WOODSMAN

We use the best British ingredients, all sourced as sustainably as possible.

We love using Wild food, and harvest much of it ourselves. All the meat on our menus is "nose to tail", so the menu is ever-changing, and we waste nothing. We even employ our head ranger to keep the deer herds on the land we manage under control and have our deer larder in the Cotswolds, where this fantastic product is processed.

HEAD CHEF GREG NEWMAN

THEY CARE ABOUT THE INGREDIENTS

Our fish comes from day boats in Cornwall.
We work with the seasons, ensuring the very best truly local vegetables are with us the day after they are picked. Our food is cooked over wood from local plantations, and charcoal coppiced less than 20 miles away.

We know the story of everything on the menu.
Our lovely staff are extremely well informed, so do ask questions about anything you're not sure of. We hope you love this restaurant as much as we do and keep coming back.

WHERE TO FIND THE WOODSMAN

www.thewodsmanrestaurant.co.uk
INSTAGRAM: thewoodsmanrestaurant

MARI CARMEM

Mari Carmen is the founder of Mari Carmen Fitness. With a background in Paediatric Nursing, she aims to empower mums to achieve fitness, health, and wellness goals.

Mari's programmes focus on mind, health, exercise, and family-friendly nutrition.

WHAT THEY ARE ABOUT

With the launch of the "Fit Mama Powered By Mari Carmen Fitness" app, health and fitness for mums are more accessible.

The app offers three programmes to cater to various budgets and backgrounds, aiming to reach moms nationwide with high-quality service.

FIND OUT MORE

Instagram - @maricarmenfitness
www.maricarmenfitness.com

SOPHIE HYAM

Sophie is a self-employed Chef running her own catering business.

A knowledgeable, competent and professional chef always going above and beyond to give her customers a dining experience they want to come back for time and time again.

Sophie shares catering premises with a local catering company and has a 5-star hygiene certification. She has a small team of chefs and waitresses she calls on as and when required, and she couldn't be busier.

Freelance award-winning Chef, Food Demonstrator, Private Dining Chef, Judge of the Great Taste Awards, Product Development Chef for Mission Foods

AWARDS

Winner: Foodie Awards Casual Dining Chef of the Year 2022
Winner: Home Cook of the Year September 2017

WHERE TO FIND SOPHIE

Instagram - @sophiehyam
www.sophiehyam.co.uk

ALL ABOUT SOPHIE

Sophie Page has run a thriving wedding cake business for over 13 years and specialises in distinctive, bold designs for couples who march to the beat of their own drum.

WHAT ELSE?

She also coaches fellow cake makers, helping to transform them from hobby bakers to becoming the stylish and confident cake artists they are meant to be. She does this through bespoke mentorship and online courses that give access to all her years of experience.

WHEN NOT BAKING

She's most often found in the kitchen with a notebook or in a restaurant tucking into dessert. She's got just as much passion for dining on great food as she has for creating it.

WHERE TO FIND

Instagram: @cakesbysophiepage
cakesbysophiepage.co.uk

MIKE SCOTT

Chef Scott is a highly skilled chef specialising in various hand-crafted chocolates. Each piece is carefully created to guarantee exceptional quality and flavour, ensuring every bite is a memorable experience.

During the interview, he expressed his passion for art, which is evident in his chocolate brand. Indulging in his chocolates feels like savouring a valuable work of art.

Freelance Chef, Food Demonstrator,
Private Fine Dining Chef,
Product Development.

WHERE TO FIND MIKE

Instagram - @scottschocs
http://www.scottschocs.com

ABOUT CHEF MARIUS

Marius is the mastermind behind The Leopard Spots, a traditional Neapolitan pizzeria. He took a break from his office job to immerse himself in the art of crafting authentic Italian pizza. Marius then purchased a truck, transformed it into a mobile pizza trailer, and devoted seven years to perfecting his craft before finally opening his first shop in 2019.

THE LEOPARD SPOTS

The Leopard Spots was launched in the middle of a pandemic and has continued to thrive since then. Marious believes that anything done with passion will be done successfully. His shop is a testament to this belief. Through hard work and delicious pizza, he has won the hearts and taste buds of his loyal customers over the years. We, at BKS, wish you every success!

WHERE TO FIND THE LEOPARD SPOTS

@theleopardspots - on Instagram
115 Regent Street - Leamington Spa

JANE DAVIS: HOMEWRECKER

Jane Davis, a Savannah native based in Leamington Spa, is on a mission to change the way y'all cheese. As an American Briton, she was shocked (read: devastated) to find no pimento cheese here.

Seven years ago, craving a taste of home, Jane waltzed into her local pub with a fresh batch and told the chef to cover her burger in the good stuff.

It's been on the menu ever since... and there, Homewrecker was born.

PIMENTO CHEESE!

Pimento cheese has been warming hearts in the southern USA for decades, so Jane became determined to make the unfamiliar familiar and convert y'all Brits from strangers to lovers of this southern classic.

"So buckle up y'all and come along for the ride; this Southern gal is about to shake up the way y'all cheese and show you how it's done where she comes from." Says Jane

WHERE TO FIND HOMEWRECKER

@homewreker_uk - on Instagram
www.homewrecker.co.uk

A GOOD CATCH

Prior to inviting Mandy to our show, I visited her restaurant and tried some of her poke bowls. I was greeted with a warm and welcoming smile. The food was delicious, and her vibrant personality made the experience even more enjoyable.

"This is my favourite lunch spot in Leamington, known for its incredible Poke boxes and delicious miso soup. It's a nutritious meal, and they offer a loyalty card for complimentary lunches. I highly recommend it; after many visits, I rate it 10/10 for exceptional hospitality and a warm welcome every time". Tripadvisor review

With 5 stars on tripadviser I could not stop my self on reading some of the reviews.

A GOOD CATCH

Mandy, who hails from Sweden, is the proud mother of two daughters. She takes immense pride in being the first person to open a poke bowl restaurant in Leamington Spa. Originating from Hawaii, this adds a unique and special touch to her establishment.

WHERE TO FIND A GOOD CATCH

Instagram - @agoodcatch
46 Regent Street - Leamington Spa

THAT GIN COMPANY

Allow me to introduce Steve, a dedicated father of two. From the moment we met, we instantly connected with him. His passion for his family is contagious, and as an entrepreneur, he truly understands the importance of service. He creates memorable experiences for his audience, not just at his Cocktail Tour bus, but also at his cocktail bar in Warwick.

Having him on the show was an incredible experience. His specialty lies in crafting cocktails tailored to each client's unique taste, which I absolutely love! Below are recipes that have been specifically created with ingredients easy for people.

To source to make at home. We had a fantastic time filming his episode, and I hope you enjoy his story as much as I do. He's truly a remarkable individual!

I picked up some fantastic tips from Steve on crafting a custom cocktail and how to enjoy it using a straw! You definitely need to check out the episode

WHERE TO FIND THAT GIN COMPANY

Instagram - @thatgincompany
www.thatgincompany.co.uk

163

WARWICKSHIRE GIN COMPANY

Let me present David Blick, the founder of The Warwickshire Gin Company based in Leamington Spa. Our paths crossed thanks to our PR representative, Amanda Chalmers. While organising my Macmillan afternoon tea, he kindly sponsored the welcome drinks for all attendees!

Before that, we filmed an episode of our show, where he shared his inspiring journey and demonstrated how to create his delicious cocktails. I have never laughed as much as we did on that filming day.

You're sure to enjoy his episode, especially since we collaborated with the wonderful Whittle's restaurant.

WHAT THEY ARE ABOUT

Using a small batch method in our Royal Leamington Spa Distillery, we distill with a traditional copper pot. While respecting traditional processes, we embrace innovation in the craft gin world.

WHERE TO FIND THE WARWICKSHIRE GIN CO.

Instagram -@thewarwickshiregincompany
https://warwickshiregincompany.co.uk/

WHITTLE'S RESTAURANT

Experience Whittle's restaurant and bistro in the stunning Grade II Listed Victorian Binswood Hall, located in Audley Binswood, Leamington Spa. The contemporary, charming restaurant features large windows and a relaxed atmosphere, just two minutes from the Regency Shopping Parade. Enjoy classical British dishes created by head chef Paul Worthy in an elegant setting, along with a more casual bistro option.

It was a pleasure to have Anja, the general manager of Binswood, share her story along with Binswood's remarkable journey. Additionally, head chef Paul Worthy delighted our audience by teaching them his traditional scone recipe during our show.

WHERE TO FIND WHITTLE'S RESTAURANT

Instagram - @Whittles_Restaurant
https://www.audleyrestaurants.co.uk

SUE CRESSMAN

WHAT THEY ARE ABOUT

This episode was full of emotions, offering inspiration and courage to all of us. Sue is a survivor of cancer twice and always with a big smile on her face.

Sue is the big boss with her husband Rick of the historic Nailcote Hall Hotel and Golf Club where she managed to raise over £500k to help local hospitals' to provide a better care to breast cancer patients creating all sorts of events at the hotel and the famous pink ball every year in October.

Our special episode focused on empowering breast cancer survivors and featured the inspiring Sue Cressman. During this occasion, I stepped into the kitchen to prepare a delicious detox oven omelette, and we enjoyed a lovely chat for the rest of the afternoon.

WHERE TO FIND NAILCOTE HALL AND SUE

Instagram - @Nailcotehallhotel
www.nailcotehall.co.uk

TAVOLA CHEF MARTIN

Welcome to Tavola – Leamington's own slice of Italy, where every dish brings the zest of Italian life, from authentic Romana pizzas to diverse pasta dishes. Tuscan executive chef, Martin Serafino infuses passion into every dish using top-tier Italian ingredients and the finest local produce. The secret to his menu, he says, is a combination of 'tradition and innovation.'

They also offer a delicious Bottomless Brunch between 11am-4pm every day.

Meeting Chef Martin from one of my favourite regions in Italy, namely Tuscany, was fabulous! Martin is such a talented Chef with an amazing heart!

WHERE TO FIND TAVOLA

Instagram - @tavolauk
www.tavaolauk.co.uk

MAGIC WINGDOM

I introduce to you Magic Wingdom, an award-winning establishment featuring wings, burgers, bao, cocktails, soft serve, and craft beer, owned by Sam Cornwall-Jones and Frankie Griffiths and she brings her chef Ash Faria. This charming, independent gourmet fast-food restaurant is nestled in a basement in Leamington Spa. Here, you'll find comfort food that beautifully blends Eastern and Western influences. While dining can get a bit messy, they provide cutlery and have kitchen roll available on every table.

We at Bia's Kitchen Show had a wonderful time learning more about the mastermind behind this incredible restaurant, as well as Chef Ash, who shared some secrets about their delicious wings. Ever wondered why they are so crispy? Now you'll find out!

AWARD WINNER

1st place: Best wild wings 2023
2nd place: Best Buffalo wing 2023

WHERE TO FIND MAGIC WINGDOM

Instagram - @themagicwingdom
www.magicwingdom.co.uk

RODIZIO RICO CHEF ANGELO STORELLI

UK's first Churrascaria, Rodizio Rico, offers authentic Brazilian flavours since 1997. The "Rodizio" style features staff rotating freshly cooked meats to your table. With locations in London, Birmingham, and Coventry, the restaurant serves not just steak but also salads, pastas, pies, fish, and Brazilian snacks. Angelo made our Christmas dinner an extraordinary experience.

We were thrilled to host Angelo, the executive chef of Rodizio Rico. In this episode, we had an additional special guest, the fabulous Kirsty Leahy, a professional radio and TV presenter. DJ and freelance host of BBC Radio, and mother to a wild toddler. Together, we explored some traditional Brazilian dishes that she had never tried before, all of which can be found at Rodizio Rico.

WHERE TO FIND KIRSTY

Instagram - @kirstyleahy

WHERE TO FIND RODIZIO RICO

Instagram - @rodizioricobirmingham
@rodizioricotheo2
@rodiziorico_cov
www.rodiziorico.com

BIA'S KITCHEN SHOW
THE TEAM THAT MAKES IT HAPPEN

This book truly embodies a journey of gratitude. I feel thankful for being able to work with something I love, merging all my knowledge and studies with my passion for empowering others. Sharing people's stories, being inspired by their lives, and serving as a conduit for positivity and good energy is truly incredible.

The key to Bia's Kitchen's success, both in the show and now the book, lies in our team, followers, friends, and the amazing guests we have. It's like baking a cake – when all these elements come together, it creates a recipe for success. Let's not overlook the music that sets the tone for each episode's opening; we are grateful to the talented Brazilian artist Zabelê and songwriters Domenico Netto Lancellotti and Alberto Continentino for letting us use their music "Nossas noites"

Most importantly, I couldn't achieve everything I do without the incredible and talented people by my side.

- Dave Perry - video and photography director
- Matt - editor and YouTube manager
- Naomi - makeup artist
- Chalmers News PR
- Lauren Perry - TikTok manager
- Bia Rodrigues-Perry - producer, director and presenter
- Katie Bell - PA

170

WHY CHOOSE MIND AS OUR CHARITY?

We're supporting **mind** for better mental health

While one in four individuals face mental health challenges, a significant number do not access the required support. This situation needs to change.

I personally struggled with postnatal depression and came close to ending my life. It was a challenging period, but I sought help and overcame it. Unfortunately, my brother-in-law didn't have the same outcome and lost his life due to mental health issues. I am determined to assist individuals like us to get the necessary support. Organisations like Mind provide valuable support, respect, and care for those in need.

WHAT MIND DOES

We advocate for mental health as a daily priority in England and Wales. We challenge the inequalities in healthcare, employment, and law that create obstacles for individuals with mental health issues.

We provide mental health support through information, advice, and local services whenever it is needed.

We unite individuals and communities who are passionate about mental health to create a powerful network and drive positive change.

IF YOU WOULD LIKE TO DONATE

Nobody should face a mental health problem alone. From a monthly donation or a gift in your will to running a marathon or getting your game on, there are so many different ways you can make a difference.

https://www.mind.org.uk/get-involved/donate-or-fundraise/

WHERE TO FIND MIND

Call our Infoline on 0300 123 3393
Email info@mind.org.uk
https://www.mind.org.uk/about-us/what-we-do/

SPONSORS

THAT GIN COMPANY

Steve Bazel That Gin Company boasts what is believed to be the largest made-to-order gin offering in the UK.

In addition to a full bar, guests at That Gin & Cocktail Bar in Warwick, have access to more than 100 unique flavour infusions – as well as a truly unique experience for their private event. And you don't have to be local to take advantage. Visitors to the website can order one of the best-selling flavours or even make up a gift set of all their favourites.

Steve has also established Cocktail Tours, after taking ownership of a new events bus! The converted double-decker can host up to 50 guests, features state-of-the-art LED lighting and sound system and a bar selling a wide range of alcoholic drinks, including cocktails/mocktails, spirits, beers, wine and fizz. It is available for almost any kind of occasion, whether it be event or private transfers; static hire at an event; exclusive on-board Cocktail Masterclass, private party or corporate takeover.

That Gin Company is also enjoying huge success with its mobile events bar bringing the party to any venue - including festivals.

Steve is incredibly passionate about his customers and ensuring they have the best possible experience and it's this reputation that has made That Gin Company synonymous with a gold standard service.

WHERE TO FIND THAT GIN CO.

Instagram - @thatgincompany
www.thatgincompany.co.uk

DAVE PERRY PHOTOGRAPHY

David is a seasoned photographer who has been in the industry for over two decades. He has studied media and film, and is the founder of Dave Perry Photography - a successful company with a portfolio of prestigious clients such as Cartier, Jamie Oliver products, Royal Mail, Co-op, and many others.

David is a devoted husband and a dedicated father of three children. He serves as the photography director and videographer for the Bia's kitchen show and this book. David has a deep passion for his work, exploring various areas of photography, including children's portraits, commercial projects, and production. He takes great joy in making people happy when they see the results of his efforts, believing that a picture should communicate its story to the viewer.

It's a pleasure to collaborate with someone so artistic and generous, as we can share our dreams and turn them into reality together.

WHERE TO FIND DAVE PERRY

Instagram - @daveperryphotography
www.daveperryphotography.co.uk

WHITTLE'S AT BINSWOOD
LEAMINGTON SPA RESTAURANT & BISTRO

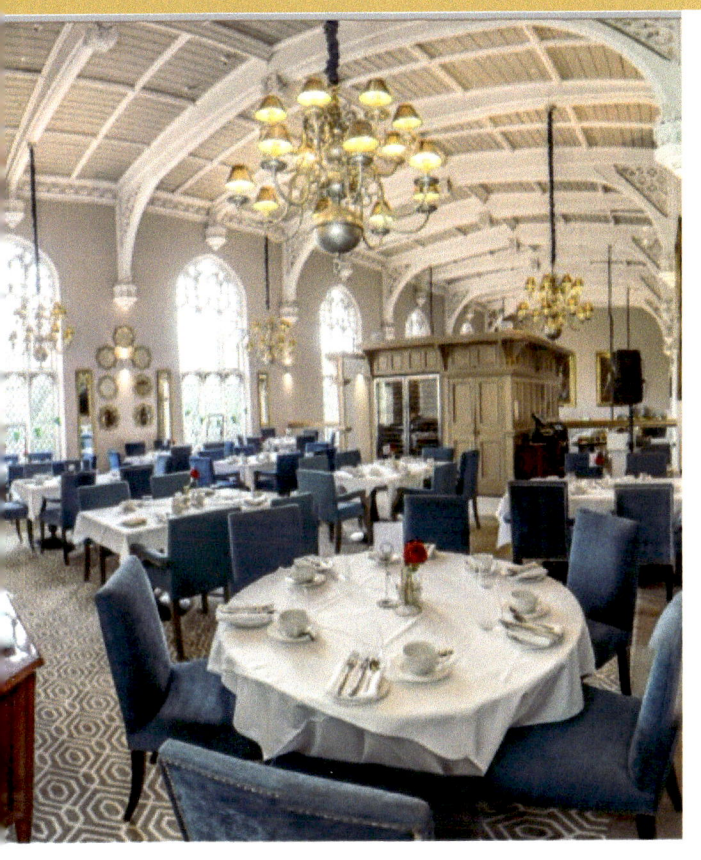

Whittle's at Binswood is a unique restaurant and bistro located in the Royal Leamington Spa conservation area, housed in the Grade II-listed Victorian Binswood Hall. The venue features a remarkable art collection and is part of the Audley Club, which includes a luxury health club and library. Whittle's charming English character and chic atmosphere offer a luxurious dining experience close to the town centre.

Built in 1847, Binswood Hall was inspired by Hampton Court Palace. Hampton Court Palace inspired its pattern brickwork, Bath and Caen stone dressings, and projecting octagonal turrets. Over the years, it has been home to several distinguished schools, notably Leamington College for Boys.

The jet engine inventor Sir Frank Whittle studied here between 1918 and 1923, and this is where Whittle's name comes from. He learned most of his engineering at his father's workshop, the Leamington Valve and Piston Ring Co. at Clinton Street, but he also touched on Binswood Hall history.

No matter your preferences or the occasion, Audley offers the ideal experience tailored just for you.

TO FIND OUT MORE

About our private function offerings at Whittle's Restaurant, please call on 01926 258 848

Whittle's at Binswood | Leamington Spa Restaurant & Bistro (audleyrestaurants.co.uk)

175

THE BOOK CHIEF®

The Book Chief is a multi-award-winning independent publishing house registered with Waterstones, offering a wide array of publishing services tailored to aspiring and established authors. Established in 2021, they specialise in self-publishing for all genres and traditional publishing for fiction authors. The Book Chief provides a friendly, open, and approachable environment, welcoming authors from around the globe. The Book Chief has built a reputation for quality and creativity with an impressive portfolio of over 300 authors from over 20 countries; they publish across diverse genres and territories, including the USA, Canada, India, Africa, South America and Europe.

In addition to publishing books, The Book Chief also manages "Authors Rising Magazine," which features author spotlights, writing tips, and insights into the publishing industry. This magazine offers inspiration and support to writers at every stage of their journey but more to help give readers some insight into the Author's journey and what motivated or inspired them to write their books. Whether you're looking to publish your first book, collaborate on a creative project, or share your expertise, The Book Chief provides comprehensive support, including flexible payment plans, to help bring your vision to life.

WHERE TO FIND THE BOOK CHIEF

44(0)7890 202375
www.thebookchief.com
sharon@thebookchief.com

THANK YOU TO...

Gratitude! My favourite word.

First and foremost, I want to extend my heartfelt thanks to God for His unwavering love, despite my imperfections, and for the incredible opportunity to write this book.

I am deeply grateful to my husband David for supporting my dreams and giving me his free time to film the show. A big thank you to my beautiful children - Lauren, Belle, and William - for their understanding during moments when I couldn't tuck them in, prepare their favourite meals, or dozed off before reading their bedtime stories.

I want to express my sincere appreciation to my family at home in Brazil: My mum Eunice for her constant prayers; my dad Leo; my sister Betânia, the best dentist ever and my nephews, Khaley and Pyetro; my brother Bruno and my nieces, Valentina, Isabel and Cibelle; and my friends for their patience while I missed birthdays and gatherings as I worked hard, all the while sending me positive energy.

I also wish to acknowledge all the incredible guests on the show who have inspired me with their remarkable stories. A special mention goes to my team: Matt, my editor, who spends countless hours ensuring everything runs smoothly; Amanda, my PR, who keeps me company during late-night sessions and makes sure my work receives the recognition it deserves; Katy, my PA, who, though new to the team, is already making a significant impact; and Naomi, my makeup artist, who helps me look my best. I also want to thank the wonderful independent clothing shops that have dressed me for the shows and awards, Sharon from NV Her and Hannah from Diffuse Retail, I appreciate you.

I'm so grateful to my dear friend and talented Brazilian artist Zabelê and songwriters Domenico Netto Lancellotti and Alberto Continentino for letting me use their music "Nossas noites" to be the show's signature tune.

Additionally, I can't overlook my publisher Sharon for her valuable advice and her designer Abu, who has beautifully captured my vision for this first book.

To my followers and subscribers, you are truly remarkable! I strive to make the world a better place, and with your support, I'm making great progress. Thank you, thank you!

Bria
07534278921